Ruppert's Modern Classic Bangerpedia 1990 - 2000

James Ruppert

FORESIGHT PUBLICATIONS

ISBN: 978-1-9160522-8-4

RUPPERT'S BANGERPEDIA 1990 - 2000

BANGERDEDICATION

To everyone who drives on the
1990s motoring margins, or aspires to.

EXCUSES EXCUSES

The Publisher and Author cannot guarantee the pin point accuracy of every single stat in this Bangerpedia, so can't be responsible for any decisions the reader makes based on the information contained inside.
Any questions?
By all means email the Bangernomics organisation for friendly free advice.
james@bangernomics.com

WHAT'S IN THE BANGERPEDIA?

THANKS VERY MUCH

Dee & Olivia
(their names are above the title after all)

Cheers to some motoring mates who sent me pictures of their '90s legends which are used on the covers.

Renault Twingo - Chris Hendrie
Fiat Coupe – Andy Sherratt
Volvo 940 – Rich Scott
Volvo C70 - Kenny Smith

All the writers and sub editors I have worked with over the years, on various publications including Parker's Price Guide, Used Car Buyer, Car Week, Car, Practical Classics, Classic Cars, Fast Ford, 911 and Porsche World, Golf Xtreme, Mini World, Mini Magazine, The Independent, The Times, The Evening Standard, Performance Car, Autocar, Free Car Mag, Bangernomics Mag and loads I've forgotten.

READ THIS FIRST

They don't sell Bangers anymore. You know, cars that eventually become cheap enough to buy with some modest savings or disposable income. They should be easy to fix, possibly cosmetically compromised, but they will do a job. That means getting you from A to B most of the time

This is the issue I now have with this latest Bangerpedia. Read the subtitle, 'Modern Classics." That means there is something of a problem when it comes to getting your hands on a vehicle from a few decades ago. Let's talk about that first.

For some of us, the 1990s were only yesterday. They were simpler, Brit Poppy times with dial up Internet. The cars were brilliant too, not just characterful and entertaining they were safer. The adoption of ABS and crumple zones, along with power steering made them easier to live with, plus they remain truly fixable, rather than being chip based lifeforms. However, when you wander around a car park these days, how many '90s survivors can you find? They seem to be disappearing fast, owned and driven only by the eccentric, or impoverished.

It is possible to drill down into the official registration stats and lose several hours of your life by discovering that there are an awful lot of models which are rather more rare than any '90s Ferrari model you care to mention. Modena officially produced just 349 F50s in 1995, but compared to a bottom of the spec sheet Citroen ZX 1.4 Avantage, well there are just two of those knocking around on the King's Highway. Meanwhile a Ford Mondeo Aspen (another base model) which peaked at 11,615 on the road in 1998, is now down to just 12 registered with the DVLA. It is possible to play this game all day long and it is immensely depressing.

So whatever happened to '90s cars? Well, twenty years plus is a long time in a car's lifecycle. Averagely, cars last a little over a decade as according to the most recent 2021 SMMT figures the median age of a scrapped car in 2015 was 13.9 years. There is always some natural wastage as cars fail MOTs and get written off of course, but also plenty of perfectly roadworthy motors were crushed. That brings us to the tragic 2009 scrappage scheme, which claimed the life of at least one Honda Integra Type R, and was underwritten by £300 million in taxpayer funds, an amount matched by the manufacturers, for vehicles first registered in the UK before 31 August 1999. As well as drumming up business for the car trade suffering after the 2008 world economic downturn, it was justified on the grounds of cleaning up the atmosphere with more environmentally friendly (their words) and safer cars. In all, 390,000 cars were taken out of circulation. Forever.

Also, the recent arrival of E10 petrol has condemned cars that can't drink it to obsolescence, pricing frugal motorists out of cars mostly made before 2002. Not everyone can afford the indulgence of super unleaded fuel. In a way then this Bangerpedia is a round up of just which '90s models you could save for posterity.

Compared to the first Bangerpedia, 'Contemporary' which put me off doing any further compilations or encyclopedic endeavours for life, this was almost fun. I liked the cars far more. They had character and real style. What we have here is most of the models offered for sale in Britain between 1990 and 2000 with yet again, one Twingo shaped exception. I avoided everything launched in 2000 as that can be found in the Contemporary Bangerpedia.

Are there overlaps with the first one, yes, there is no avoiding that, but many of cars have been re-rated, the pictures may have changed and some of the opinion. The simple truth is that an awful lot of the vehicles from '90s are not Bangers. As before there are plenty of vehicles which will never be Bangers short of an economic and societal breakdown. So Astons, Lamborghinis, Ferraris and the high end Porsches are mostly outside of the remit. Anything else I've just forgotten.

What we do have here is some vehicles which are most definitely classics and inevitably will have gone up in value. So what's the point of this Bangerpedia? I don't know really. You can go online and get just about every fascinating fact and multiple images you could ever want. I believe what

I'm doing is identifying what were Bangers and what you could have bought ten years after they were made, for less than you would imagine. Slightly pointless then, but no prices are listed so you can't even dream. Double pointless. I believe this is worth doing, it isn't a searchable online database, it's a book that you can refer to, argue with, maybe reminisce and keep in the smallest room on the house.

Some of the listings might surprise you, there is the odd Roller, 8 Series BMW and complicated Mercedes, but at one time in their lives, would have been relative Banger money, so that's why they are here. Possibly with a dire warning or two.

If I have missed anything, by all means let me know, I might care and possibly include it in the future, with a credit to you. This does though have my name on it, so this is my angle on the whole encyclopedia and what is and isn't a Banger. If you don't like it go and write your own. There is a lot for anoraks to argue about, so be my guest.

Like the Contemporary Bangerpedia, this was a difficult bit of literature to assemble. I tried to use printed sources, you know books and magazines, as I have a lot of that real stuff on shelves going back many years. Some of it I actually wrote. Indeed, I was part of the team that put together the Parker's Chooser magazine twenty years ago and I did compile and edit Car's GBU (Good Bad Ugly) section for a few years.

I used old books, magazines, some brochures, those wonderful motor show guides and the odd website. Incredibly that did not necessarily unravel the mystery of what the literage of a random '90s boot areas were with the seats up and folded.

What I did when it came to many of the variable figures was put a spread of relevant data, be it official combined MPG, or MPH. That's because when people search for used cars they need to know what the best and worst case scenario is. Not only that, the information varies slightly as a 4×4 buyer might want to know the maximum braked towing weight rather than the top speed or acceleration. The way buying used cars actually works is that you often end up with something you quite possibly had not even considered. So don't use this as a definitive reference work. I've actually deleted most of the CO2 entries as they become irrelevant, free up some space, I disagree with them and it allows me to do some random things like the turning circle and on one occasion 'numbers built' which may become a feature of future Bangerpedias. Indeed, I might have deleted the number of doors or seats, when it is bloody obvious, so that the entry will fit the format. The Bangerpedia is really me just doing what I feel like when it is the early hours of the morning.

It is imperfect and data expert Keith WR Jones is currently still laughing about all the inaccuracies in the Contemporary Bangerpedia, so this Modern Classic edition will give him a few more years entertainment.

I have stuck with the black and white images with a touch of retro art effect which went down well enough to act as a recognition guide of sorts. Like before, there are a minimal amount of 'what to look fors'. This is done on the basis that, 'they don't all do that' as some might, others might not. Every used car is different. That's because it may not have been serviced, ever, could have been in an accident, or done loads of short stop/start journeys that may be slowly killing the engine. When it comes to mechanicals, I'll only take advice from Mark, Ian, Matthew Ted, Rob and Kenny because they are actually qualified to tell me off. If your best mate's girlfriend's mother had a Lancia Dedra that did two million miles on its original shock absorbers, I don't actually care.

I'm sure you have got the message that this is my Bangerpedia. My rules and mistakes. Rather like I believe it is possible for a convertible, sports car and luxury barge to be a Banger, back in the '90s. These were affordable cars which could be any shape, size or make and you will find them all right here.

So far my attempt to turn Bangerpedia into a real life game have not gone well. I expected the streets of our great nation to be full of people carrying Bangerpedias, elbowing each other out of the way to tick off the Bangers they spot. That does not seem to have happened, but I live in hope. I think that the arrival of the Modern Classics edition could broaden the appeal and give this pseudo Olympic sport a real boost. I will keep you posted as to its progress.

Like before, I suppose if you could prove you had essentially I-Spyed all the motors in this Modern Classics Bangerpedia (car shows allowed) and have some easy way to prove it I'll put your name in future issues and maybe think of some prize or something.

I hope the Modern Classics Bangerpedia proves to a bit useful rather than being a definitive guide to all the vehicles produced over 1990 to 2000, because it isn't. Where it could come in handy is as a choc on one of the wheels when you inevitably have to jack your Banger up. Mostly though I think it will just start arguments between used car clever dicks. Provided they have bought a copy I don't mind. I do feel the need to restate that I accept no responsibility for any decision you make based on the contents of this Bangerpedia.

Incidentally, there will be a 1980s Bangerpedia, provisionally titled, Practical Classics. I now feel a strange compulsion to log all the post war cars sold in Britain that at one point in their existence qualified as Bangers. Let's see if I can keep going.

With all that in mind, always remember that buying any used car, whether or not you have a Bangerpedia stuffed up your jumper, is much a matter of luck as well as judgement. That's why I always remind used car buyers to always beware of the dog.

James Ruppert...
...at various points in 2022, which turned out to be the worst year for breakdowns and old car related trauma in recorded history, but still in Bangerfolkshire, The Great Banger Isles.

Bangerpedia Decoded

1990 - 1999 are the build years although sometimes there is a disparity between the on sale, or launch, production and registration years. Overall I have triangulated the data to arrive at the best possible beginning and end dates based on the available facts. Depending on models I have included some that were made from 1991 and ignored some others, because of space and over tiredness, or I just thought, I'll cover those in the next one.

Engines: Engine sizes for petrol and a D means diesel. **CO2:** Spread of g/km for taxation purposes, most have been deleted or ignored as by this age it is not relevant **MPG**: Best to worst official combined figures **BHP**: From time to time I might have squeezed in the old fashioned brake horse power. **Gearbox:** Man/Auto **Boot:** Litres for where relevant seats up and folded **Seats**: Number of, sometimes 2+2 when there is barely enough room **Doors:** Number of, unless it was so obvious, I've deleted for space. **Towing:** Maximum braked in kg **Top Speed**: Fastest and Slowest, not always the turbo-nutter one **0-60:** Best and worst in Seconds, it could actually be 0-62 because of kph element, but thought that the 60 is neater unless indicated otherwise. **Weight**: Kerb unladen in kgs. **Turning**: The circle the banger will turn in metres.

Then there is the rating system. This involves Slog the Bangernomic Dog who has been on this journey for some time. It used to be beware of the dog, but in the interests of clarity it has to be the more Slogs the better. Readers would associate it with a star system anyway.

Just the one Slog means you should definitely avoid. That seems obvious. This is probably a triple whammy of potential unreliability, compromised fixability and the overall expense of doing so. The truth is that many upmarket late model cars should be avoided, but might have other saving graces, like it is nice to drive. Or just does what it is designed to do, like off road.

Two Slogs is a fairly questionable Banger. Quite likely to have an average reputation for not breaking down. It might have some added element of overall averageness, be it in the practicality, driving or repair departments.

Three Slogs is most Bangers. It averages out as being good value for money, easy to own, does what it originally said in the manufacturers brochure. This is the default rating. Every Banger starts with a three and we work up or down from there.

Four Slogs is jolly good going for any Banger. If you've got any standards then this is the one to aim for, even though as we know buying a Banger is not a precise science. You end up with what you can afford or what is available. If it is a four star Banger then great.,This will do a job with some style.

Five Slogs is the Banger Gold standard. These are going to be the real legends which are unarguably brilliant Bangers. Most of these are going to be Japanese, but they will have a genuine reputation for being very good indeed at being a Banger. I seem to have given an awful lot of 5s this time round, clearly an a prime Banger decade.

No Slogs? I think they all score at least one. I might have given one Banger a 6 Slog rating for some reason,

Anyway, here they are. Bangers grouped together according to their dimensions and/or their claimed abilities and attributes.

Miniscule - Teeny Tiny Town Tots

Here is a brand new category just to confuse the small number of Bangerpedia aficionados. This is a recognition that before the so-called supermini, there was just the mini, hence Miniscules.

Yes, the '90s was the last brilliant gasp of the actual City Car. This was the last decade that the original small wonder, in the shape of the mini, could still be bought brand new. It was responsible for creating this class, although it was properly outclassed by then and my Slog rating reflects this. There is also a good gaggle of Fiats and their related spin offs which shows how strong they were in this market.

Some manufacturers were also keen on trying something new, hence the great leap forward in minimal car design courtesy of the Swatch related Smart. Mercedes made it better but never made any money from it, which is how at the time of writing the Bangerpedia the smallish car is now dead. More on that in a few pages time.

The cheapest way to make small cars is to keep on remaking old ones, sometimes with different badges like SEAT, or using cheap labour in pursuit of hard western currency like Zastava. All they did was make a lowest common denominator motor at the lowest possible price.

There were always basic transportation solutions, but that involves going down the weird world of mainly French Quadricycles which Reliant did import for a while. Also there was the Piaggio Ape scooter vans and far eastern Tuk Tuks. Instead I've stuck to proper tiny cars with the correct compliment of wheels.

These Miniscules are the purest of motor cars. Anyone can build a V16 supercar, that does VTOL and has a delicate undies spin cycle. It is a work of genius to keep things so simple, yet practical and economical. It took one brilliant man, Sir Alec Issigonis, to create the mini. A great designer, like Giorgetto Giugiaro was responsible for the cool simplicity of the Fiat Panda.

Make arrangements to buy and enjoy one while you still can. Minis can be stupidly expensive, but some Pandas and most Twingos aren't.

Fiat Cinquecento 1991- 1998

This box is brilliant around town, refined and bags of fun. Sporting is noisy fun and the most desirable. Basic S is best whereas SX has glass sunroof, remote locking and electric windows. Old school Polski build quality never the best. Head gasket on the 1.1 Fire engine, clutch, rust in fuel tank, gearboxes weak. Now super scruffy.

Engines: 900/1.1 **BHP**: 41- 54 **MPH**: 87-93 **MPG**: 45.3-47.6
Gearbox: Man **Boot:** 170/810 **Seats**: 4 **Doors:** 3

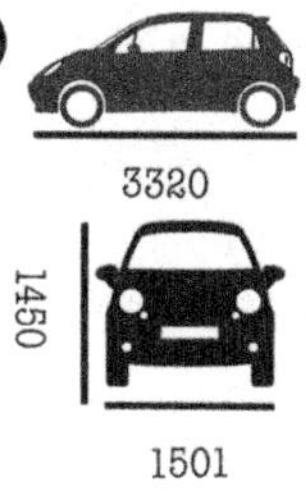

Fiat Seicento 1998 - 2004

Fiat are good at tiny cars and although this seems quite ancient it does what a city car should do and soak up pot holes, provided the suspension is OK. Won't notice the lack of power steering. It is a Fiat so expect random failures. High spec SX likely to have a lot less working (windows, boot release). Quite an old piece of kit now. There will be rust.

Engines: 1.1 **CO2:** 143-155g/km **MPG**: 43.5-47.1
Gearbox: Man **Boot:** 170/810 **Seats**: 4 **Doors:** 3

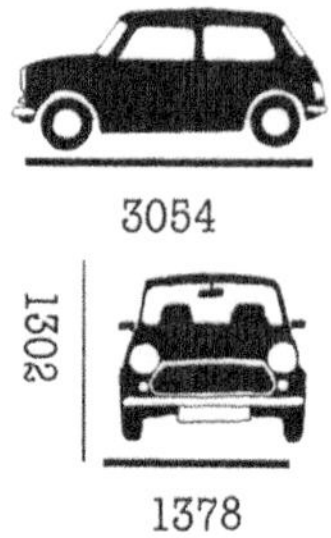

Fiat 126 BIS 1987 - 1992

Bigger than a mini? Frantically noisy 2 cylinder with a rock hard ride and barely enough room for two. Shoddily Polski put together and not a lot going for it but quirky charm. Mostly a rust bucket these days, not much electrical to go wrong, engine has a finite life, so big miles without a rebuild spell trouble. Lots of oil leaks and gearbox not the strongest.

Engine: 704 **MPG**: 50.4 **BHP**: 26 **Turning**: 9.3m
Gearbox: Man **Boot:** 110 **Seats**: 4 **Doors:** 3

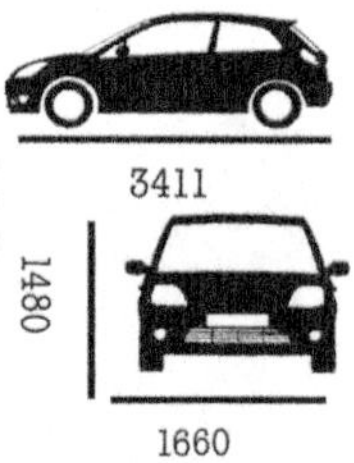

Fiat Panda 1986 - 2002

Still going strong a few decades ago. This Mark 2 ran from 1986 onwards and there was a ton of models to choose from. Charming, but often in a badly finished, falling apart sort of way. 1000 is the pick of engine bunch, CVT might not work anymore and the 4×4 is sadly very rare indeed. Rust, gear selection, bore wear blue smoke.

Engines: 800 • 900 • 1.0 • 1.1 **CO2:** 152-160g/km **MPG**: 40.9-48.6 **Gearbox:** 5spd **Boot:** 269 **Seats**: 4 **Doors:** 3

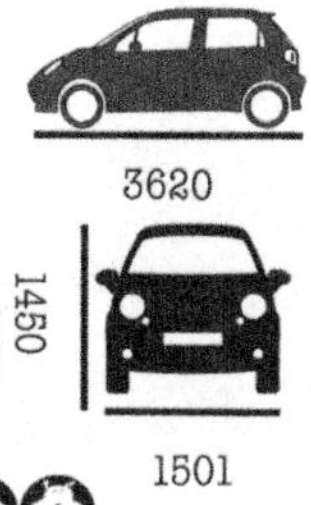

Ford Ka 1996 - 2008

Well this is a legendary little thing which drives like a demon but isn't fast enough because of the ancient iron engine to cause any real damage. Rust is the big killer and the fading plastic bumper just looks sad, but makes it even more lovable. Lots on sale, noisy and under the bonnet a bit tight. Despite rust issues what a Banger should be. Cheap to fix.

Engines: 1.3 **CO2:** 163 g/km **MPG**: 42.2 **Gearbox:** Man
Boot: 186/724 **Seats**: 4 **Doors:** 3

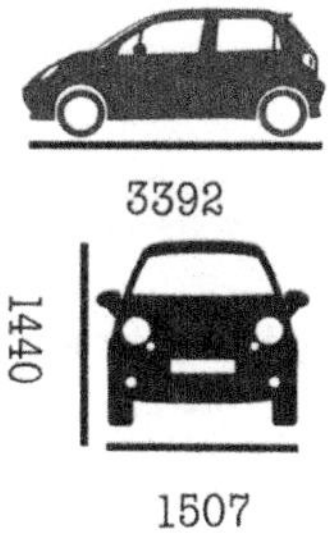

Lancia Y10 1985 - 1991

Upmarket Italian box which was reassuringly quirky. The teeny Turbo was a hoot, but most buyers went for the 1.05 litre Touring then the 999cc FIRE. Most spelt trouble of course, but there were special editions, extra sound deadening and suspension upgrades in '89. CVT Selectronic worth avoiding. '90 Series 2 available in UK for 5 minutes.

Engines: 1.0 • 1.05 • 1.1 • 1.3 **MPG**: 34 - 52 **Gearbox:** Man
Turning: 9.8 **Weight:** 720 **Boot:** 195 **Seats**: 4 **Doors:** 3

MCC Smart City Coupe 1998 - 2003

In the beginning there was the MCC one. A Swatch on four wheels and it will always look like it comes from a utopian future. The originals came to the UK as grey imports and the left hooker layout meant you stepped onto the pavement and that makes it the best City car ever. Plastic panels and scaffolding grade safety cage. ECU failure kills the car.
Engines: 698cc **CO2:** 118g/km **MPG**: 57.6 **Gearbox:** SemiAuto **Boot:** 150 **Seats**: 2 **Doors:** 2

Renault Twingo 1993 - 2007

Euro only, but we love it, see cover. Drivers will be sitting on the wrong side which makes it cooler. 1.2 engines did the hard work and there was the choice of an 8V (60hp) or a 16V (75hp). Has all the basic charm of a Renault 4. Got styling makeovers in '98, '00 and '04. Worth rescuing from abject neglect on the continent. Timing belt changes needed.
Engines: 1.2 **CO2:** 138 g/km **MPG**: 43.5 **Gearbox:** Man **Boot:** 168/261 **Seats**: 4 **Doors:** 3

Rover Mini 1989 - 2000

Two flavours: cooking mini and Cooper, 'properly seats four and remains the most fun tiny car evah. Everything else more comfy and easier to put the shop in. So still a heritage rather than sensible buy. 96 safety upgrades, air bags and side bars. As rusty as a '70s from the factory, blown head gaskets, collapsed suspension, rattly timing chain. **Engines:** 998 • 1275 **MPG**: 40.1-49.1 **BHP**: 41-63 **MPH**: 80-90 **Turning**: 8.9 **Gearbox:** Man **Boot:** 120 **Seats**: 4

SEAT Marbella 1988 - 1995

Spanish Panda, just a very old fashioned holiday rental. Crude with back to basics appeal, loadsa special eds, GLX had rear wash/wipe and cloth trim, so utter luxury. Shabby paint, sun bleached plastic, oil leaks, snapped clutch cable, worn suspension, wheel bearings, shabby and neglected because it isn't a Fiat. **Engines:** 843 • 903 **Turning:**9.6 **MPG**: 42-44 **MPH**: 78-83 **Gearbox:** Man **Boot:** 269 **Seats**: 4

SEAT Arosa 1997 - 2004

A down market and duller VW Lupo. It is a tiny little thing only put toddlers in the back. S and SE trim will give you things like folding rear seats, ABS brakes and fairly pointless central locking. Diesels are an option, 1.0 is fine for buzzing around 1.4 offers more, especially as a Sport. Facelift 2001. Failing oil pumps, seized brake calipers.

Engines: 1.0 • 1.4 • 1.4D **CO2:** 119-180g/km **MPG**: 37.7- 64.2 **Gearbox:** 5spd/Auto **Boot:** 130/790 **Seats**: 4 **Doors:** 3

Volkswagen Lupo 1999 - 2005

A high quality small car, a VW Phaeton but with less doors and a lot less bother to own. The diesel does a million miles to the gallon, the Eco 3L (78mpg) and the 1.4 is shockingly perky, GTI though is one of those appreciating classics. This is a very classy little thing which only just about takes a bag of shopping in the boot, if that matters. Apart from some electrics. Basic SDI diesel excellent.

Engines: 1.0 • 1.4 • 1.4D • 1.7D **CO2:** 119-168g/km **MPG**: 40.4-64.2 **Gearbox:** Man **Boot:** 130/830 **Seats**: 4

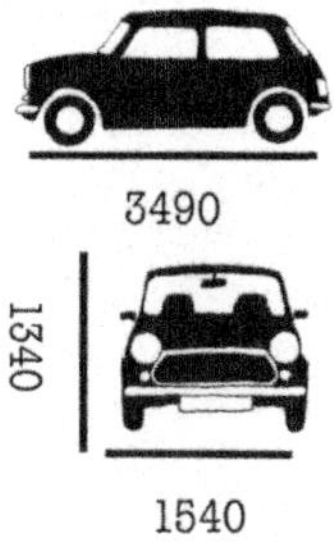

Zastava Yugo 45 •55 •65 1986 - 1993

A fairly grim and built to a price sort of thing when brand new. Back of the lot used orphan, or a one private owner example rotting in the drive. Neglected and a problem to get parts for, retro Eastern Bloc cool now, otherwise marginal with the few survivors having everything wrong with them. **Engines:** 903 • 1.3 **BHP**: 45•65 **MPG**: 38 - 45 **Gearbox:** Man **Weight:** 750 **Boot:** 170/750 **Seats**: 4 **Doors:** 3

Super Shoppers - Bangerminis for Errands

These are called superminis. Not by the people who buy and run them, that's what the car industry marketing morons decided to name them. Nope, our soopers are shopping vehicles, school runners, commuters, but in the cheapest and most practical package. These have now been a part of the motoring landscape for 20 years and these are some of the best and worst that you can buy.

Perversely I decided to include three wheelers which are not technically cars. They are also not remotely super, although there was a Supervan version which I may have to include in some future Bangervanpedia possibly coming out in 2035. No, the Reliants are in because they are part of British motoring culture. We all knew someone who was a biker by summer, but in winter it was usually his Mrs who wanted a roof on their chosen transportation rather than a sidecar to squeeze into, (see Olive from On the Buses).

There are some proper old supermini legends here. The Volkswagen Polo really came into its own and just got much better, if less basic. The Volkswagen Group fully understood this sector. Most desirable is the Audi A2 which was misunderstood by everyone and remains one high tech direction that smaller cars could have developed further. Meanwhile, VW's cross branding extravaganza gave us the option of getting a Polo with a Skoda badge on it, or a SEAT one. Value for money is really what these shoppers are about and the Fabia and Ibiza offered so much to buyers.

More Far Eastern excellence courtesy of Nissan's Micras and Toyota's Starlet. Dynamically terrible according to motoring hacks who missed the whole point of them entirely when all the owners wanted was a car that would not break down. Not only that, there was a generation of drivers who owe their licence to the safety and steady progress that they delivered which enabled the learner to complete a three point turn and emergency stop.

Ford Fiestas were great. What a tragedy that these have to die on economic and ideological grounds. The motor industry has abandoned its core customers and clearly has gone mad. Enjoy this pre-owned while you still can.

Audi A2 1999 - 2005

It looked like the future or at the very least like a very posh van with the ultra lightweight Audi space frame. Using aluminum made it very light at under a ton, but pricey to fix if bent. Old factory radios rubbish and diesel turbo could be blown. 1.4 petrol or diesel is pretty economical. Quite possibly the classiest small Banger ever built.

Engines: 1.4 • 1.6 • 1.4D **CO2:** 116-144g/km **MPG**: 46.3-64.2
Gearbox: Man/Auto **Boot:** 390/1085 **Seats**: 4/5 **Doors:** 5

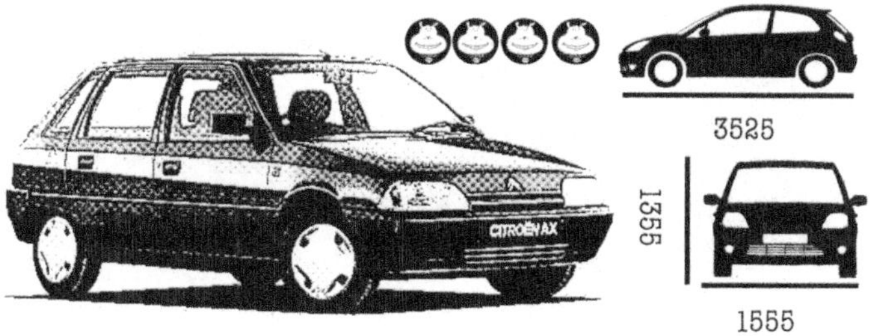

Citroen AX 1987 - 1997

The very definition of a lightweight hatch, folded cardboard construction means some spectacular mpg figures. Endless special editions, Pug 106 related, but better use of space, comfortable, just rear legroom marginal. Small petrols get smoky and oil leaky with age. Easily dented and damaged. 1.4 very nippy, easy enough to look after.

Engines: 1.0 • 1.1 • 1.3 • 1.4 • 1.6 • 1.4D/1.5D **MPG**: 41.2-61.7 **Gearbox:** Man **Boot:** 273/668 **Seats**: 5 **Doors:** 3/5

Citroen Saxo 1996 - 2003

Those after the full zero spec, one wheel trim missing Francophile experience (pedals set to the left) will find it in a sexy Saxo. Huge fun to drive, 1.1 First has less than nothing. VTR for when you stray out of town. Often modified very badly, clutches go bang and neglect blows head gaskets. Better built from 2000.

Engines: 1.1 • 1.4 • 1.6 **CO_2:** 145-159 g/km **MPG**: 38.7-46.3 **Gearbox:** Man/Auto **Boot:** 280/953 **Seats**: 4 **Doors:** 3/5

Daewoo Matiz 1998 - 2005

Interesting styling (Giugiaro who did the Golf) which resembles a funky delivery van. Actually quite nippy and easy to live with. The three cylinder engine makes a quite a roarty noise. Seems to have niggly engine issues once it is a few decades old. SE+ model does not add much to the mix. Then became a Chevy see back. Can be neglected and tired.

Engines: 800cc **CO_2:** 149 g/km **MPG**: 46.3 **Gearbox:** Man/Auto **Boot:** 167/624 **Seats**: 4 **Doors:** 5

Daihatsu Cuore 1997 - 2003

Not great to look at or drive, but it will last pretty much forever, if not longer. Another three cylinder thrasher that is truly at home in the City, and a bit all at sea elsewhere. Rust is the biggest killer now and finding parts is an ever bigger issue which starts to make insurance and just keeping it running more of a challenge.

Engines: 1.0 **CO2:** 127 g/km **MPG**: 53.3 **Gearbox:** Man/Auto
Boot: 155/428 **Seats**: 4 **Doors:** 5

Daihatsu Charade 1987 - 1993

Great little thing and for a small car at the time, power steering and automatic on the options list. Forgotten GTi is certainly fun, reasonable room for four inside. Reliability legendary for a tiddler and any survivors just keep on going. Parts will always be the issue and it is tyres, exhaust, brakes and suspension that are worn out. Not a charade.

Engines: 1.0 • 1.3 **MPG**: 47.9-58.9 **Gearbox:** Man/Auto
Boot: 152/384 **Seats**: 5 **Turning**: 9 **Doors:** 5 **Weight:** 720

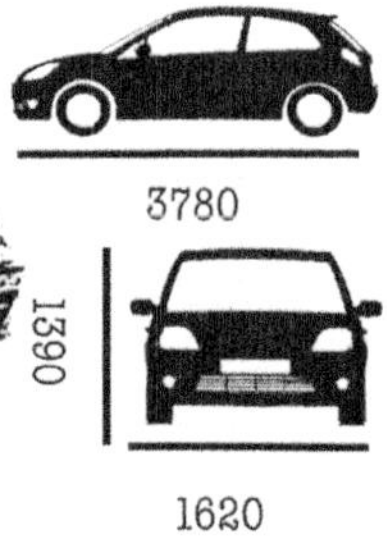

Daihatsu Charade 1993 - 2000

Dullest Charade of all, very Fiesta-like. Not a lot of equipment, Xi delivers power steering, GLXi gives you some air. Bit sharper (engine and suspension mods and safer from '96 with airbags. Loads of special editions. 1.5s are quick and there's a GTi. Tough but the airbag was prone to going off in the old days.

Engines: 1.3 • 1.5 • 1.6 **MPG**: 39.7-42.1 **BHP**: 84-105
Gearbox: Man/Auto **Boot:** 152/384 **Seats**: 5 **Doors:** 3/5

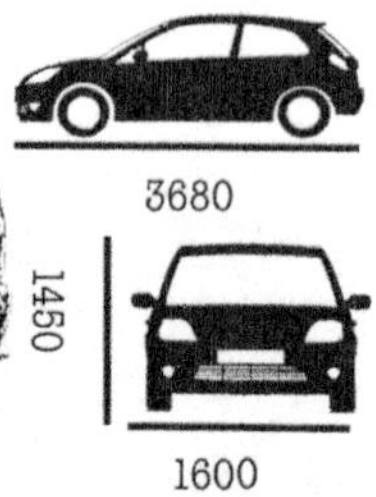

Daihatsu Sirion 1998 - 2004

Looked like an over chromed grey import (A.K.A. Doria/Boon and Toyota Duet/Passo), a posh Charade and the 1.3 is ex Yaris, so reliability never an issue. 3 cylinder 1.0 otherwise with a decent spec beyond the basic E, otherwise you have air con and eventually side airbags. Owner neglect damages suspension and coolant leaks.

Engines: 1.0 • 1.3 **CO2:** 129-134g/km **MPG**: 49.6-51.4
Gearbox: Man/Automatic **Boot:** 235/890 **Seats**: 5 **Doors:** 5

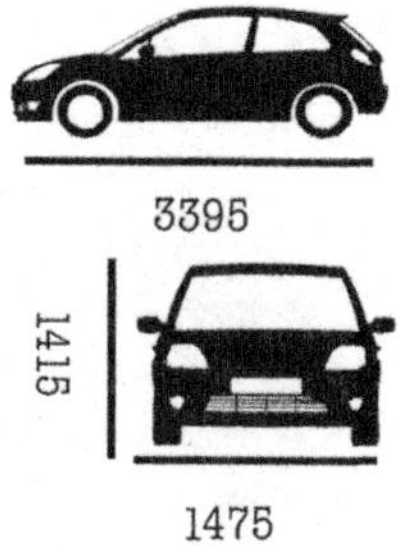

Daihatsu Mira 1993 - 1995

If you are interested this is the L200/201 model badged as Cuore, Domino, Handivan and Ceria elsewhere this is a hyper rare box. Tiny 847cc engine so economical and buzzy. Long forgotten and there are interesting micro Kei versions, there must be a prize now if you can find one and unearth the parts to fix it.

Engines: 847 **BHP:** 69 **MPG**: 50 **Gearbox:** Man/Auto **Boot:** 160/414 **Seats**: 4 **Doors:** 3/5

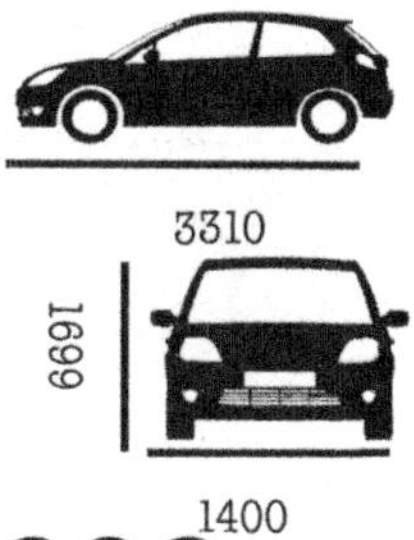

Daihatsu Move 1997 - 1999

Short lived van alike hatch was a funny little thing which was a surprisingly accommodating five seater with loads of headroom. There was a Move+ which had a rev counter and radio, big deal. Not great handling, or that fast, but that's not the point. Mechanically tough, but not looked after so lots expired early through overwork and under maintenance.

Engines: 850 **BHP:** 42 **MPG**: 47.1 **Gearbox:** Man/Auto
Boot: 316/1323 **Seats**: 4 **Doors:** 5

Fiat Uno 1983 - 1995

Wonderfully exuberant small family hatch that is great to drive and massively practical. Superb inside, good seats and very easy to live with with excellent economy with the tiny engines and diesel. There always were ownership issues, oil leaks, clattery tappets, gearbox bearings, short clutch life, CVT which is pricey, plus electricals and leaky driveshafts.

Engines: 903 • 999 • 1.1 • 1.4 • 1.7d **BHP:** 45-118 **MPG**: 36-51
Gearbox: Man/CVT **Boot:** 248/968 **Seats**: 5 **Doors:** 3/5

Fiat Punto 1999 - 2003

The snazziest looking hatch on the block. Can be flakey when it comes to using as a daily. 1.2 petrol is the basic choice, either 8V or 16V and there are diesels. Big old boot, HLX is the air con spec and ELX has a sunroof, which is cademic if it doesn't start. Original has bags of character and possibly rust. Head gasket failure.

Engines: 1.2 • 1.8 • 1.9D **CO2:** 136-197g/km **MPG**: 34-57.6
Gearbox: Man/Auto **Boot:** 264/1080 **Seats**: 5 **Doors:** 3/5

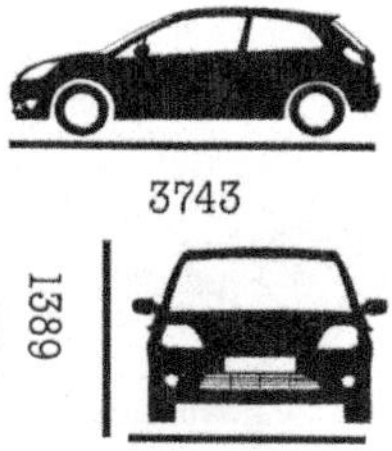

Ford Fiesta 1989 - 1997

Fiesta grows up with a five door version. Bigger than before and became available with ABS. Despite all those fancy Quartz special editions and well equipped Ghias this remains one of the toughest and best superminis ever. Still has oil leaks, heads gaskets, coolant leaks, shot gearbox mountings and clutch adjusters. Classic lived on alongside Mk 4.

Engines: 1.0 • 1.1 • 1.3 • 1.4 • 1.6 1.8D **BHP:** 45-110 **MPG**: 38.7-53.3 **Gearbox:** Man/Auto **Boot:** 311/1093 **Seats**: 5

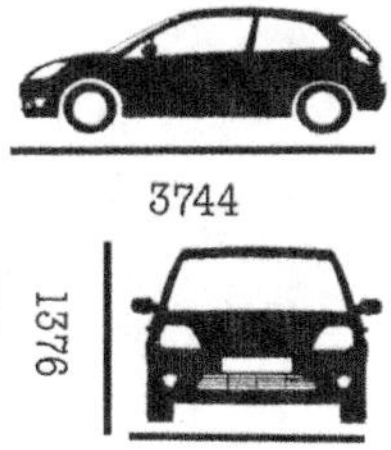

Ford Fiesta 1995 - 1999

Some say the oval grille is iconic, it is simply Guppyish, Fitted with electric door mirrors, and windows, most significant was the Zetecs which combined with revised suspension and steering made it a hoot. . Worn suspension bushes, noisy tappets, maybe a broken gearbox. Largely unloved, underrated, overlooked.

Engines: 1.25 • 1.3 • 1.4 • 1.4D • 1.8D **MPG**: 38.9-47.9
Gearbox: Man/Auto **Boot:** 311/1093 **Seats**: 5 **Doors:** 3/5

Ford Fiesta 1999 - 2002

The post fish face one is the Mark 5. The 1.25 Zetec still does the hard work and will keep going forever but now there is the luxurious 1.6. Ghia and Zetec-S. Good to drive, not that spacious, but it is cheap and fixable, which is all the committed Bangerthusiast needs. Worn suspension bushes, noisy tappets, broken gearbox. Just over tired by now.
Engines: 1.25 • 1.4 • 1.6 • 1.4D • 1.6D **CO2:** 144-171g/km
MPG: 38.7-53.3 **Gearbox:** Man/Auto **Boot:** 250/929 **Seats**: 5

Kia Pride 1991 - 2003

Regurgitated '80s Mazda 121, out of date when introduced, but not entirely charmless. Cheap and economical. The old tech is utterly bomb proof. Safety issues for some until side beams in '98. Fuel injection and a bit more power for '94. LX has all the toys like a locking filler cap. Sunroofs can leak, CV joints a weak spot, otherwise granite. **Engines:** 1.1 • 1.3
MPG: 41.9-44.4 **Gearbox:** Man/Auto **Boot:** 205/510
Turning: 8.7-9.4 **Seats**: 5 **Weight**:795+ **Doors:** 3/5

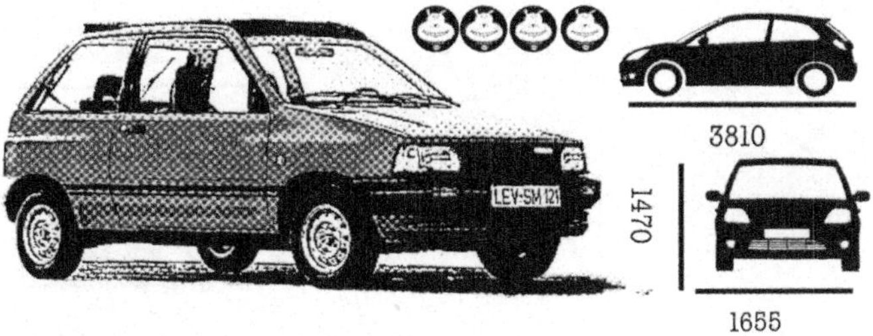

Mazda 121 1988 -1991

Here's the pre Pride version which was not just a hatch, but also available in saloon format... .Oh yes and canvas roof was option. Not a lot of specification, but the rear seats would slide to create space. Really basic but reliable. Pretty much extinct, not cherished just old and tired. So just check brakes, exhaust and consumables...

Engines: 1.1 • 1.3 **BHP**:54-65 **MPG**: 41.9-44.4 **Turning**: 9 **Gearbox:** Man/Auto **Boot:** 205/510 **Seats**: 5 **Doors:** 5

Mazda 121 1991 - 1996

A very funky bubble styled saloon that could also be ordered with a roll up roof, called a 'Cabrio Top'. Pretty fast but the engine is noisy and the automatic only gearbox made life easy enough even if the lack of power steering didn't around town. Chic enough inside too and extremely well built, last of the characterful small Mazzers before they became Fiestas,

Engines: 1.3 **BHP**:72 **MPG**: 49.6 **Turning**: 9 **Gearbox:** Auto **Boot:** 290 **Seats**: 5 **Doors:** 4 **Weight**: 815

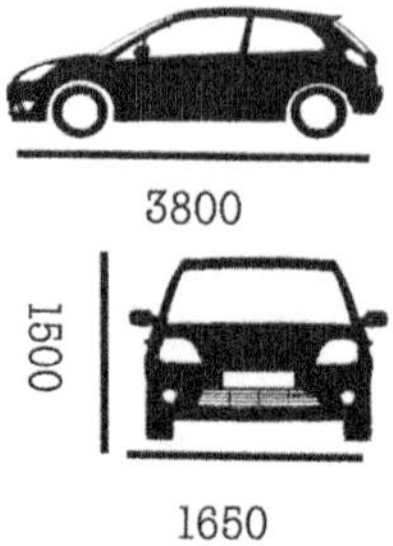

Mazda 121 1996 - 1999

Yes it is a Fiesta but with a Mazda badge. Standard power steering, expensive new, proper value used. A great sporty drive, refined and decent reliability and fixability. Brit built of course, ZXi has sunroof, tints, central locking, alarm, airbag and split rear seat. 5 door from '97. Worn suspension bushes, noisy tappets, maybe a broken gearbox.

Engines: 1.25 • 1.3 • 1.8D **MPG**: 42.2-47.9 **Turning**: 9
Gearbox: Man/CVT **Boot:** 311/1093 **Seats**: 5 **Doors:** 3/5

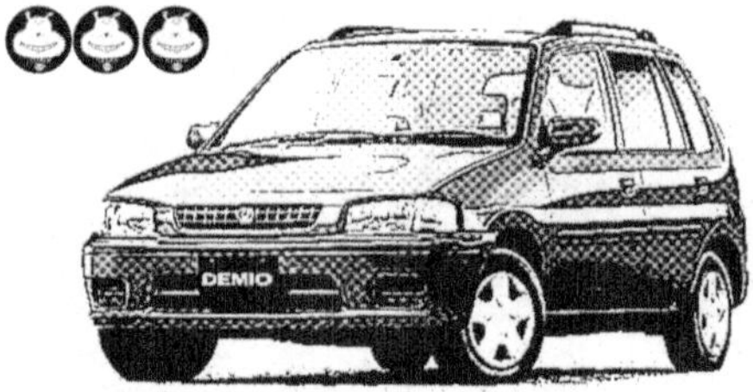

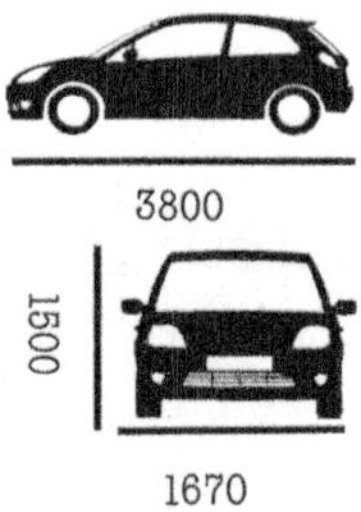

Mazda Demio 1998 - 2003

More of a maxi hatch in more ways than one. Replaced the long forgotten 121 as a roof railed micro MPV. The party trick though was reclining the folding seats into a cheeky double, or a flat bed van that could swallow almost anything. Enthusiastic 1.3 engine. No weak points, just the fact these can be old and tired. So just check brakes, exhaust etc.

Engines: 1.3 • 1.5 **CO2:** 169-176g/km **MPG**: 38.7-40.3
Gearbox: Man/Auto **Boot:** 280/950 **Seats**: 5 **Doors:** 5

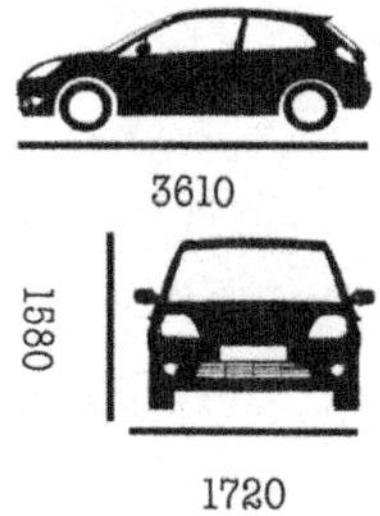

Mercedes A - Class 1998 - 2004

Supermini with a premium badge. A modern people/luggage packaging masterclass, that's a doddle to park. Sliding and removable seats, so it's a van too. Good petrols, quick 1.9, 2.1, but 1.4 is fine runs on the rock hard suspension. Stretched 'L 'ones from 2001. Diesel glow plugs trouble, water leaks, suspension bushes, power steering failure.
Engines: 1.4 • 1.6 • 1.9 • 1.7D **CO2:** 131-184g/km **MPG**: 36.9-57.9 **Gearbox:** Man/Auto **Boot:** 350/1190 **Seats**: 5 **Doors:** 5

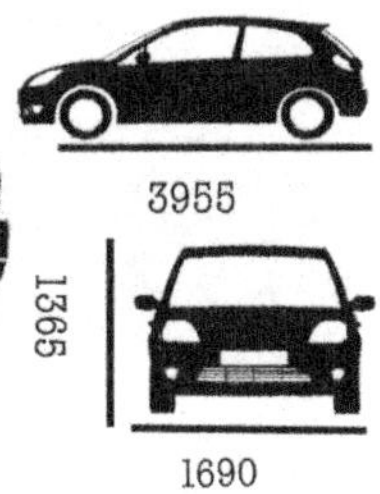

Mitsubishi Colt 1990 - 2004

Ancient old thing that goes back to the '80s. In '92 it was refined and restyled. There's a load of engines from the shopping 1.3, mid range 1.5 and hot-ish 1.6 and GTi 1.8. Blimey there were updates every five minutes. There was a clever automatic gearbox. The only model you remember is the sporty Mirage. Rust and neglect, brakes and exhaust.
Engines: 1.3 • 1.6 **CO2:** 156-171g/km **MPG**: 39.2-42.8 **Gearbox:** Man/Auto **Boot:** 240/830 **Seats**: 4 **Doors:** 3

Nissan Micra 1982 - 1992

Rumour is that this is a folded paper ITAL design like a Japanese Polo and just as good. Badged as a March bit ecame a lot of other fancy things, but in the UK it was the best of the new minis and taught a generation to drive. Rattly timing chains. Rusty bodywork and suspension obviously. Mass airflow sensors and starting issues. Mostly extinct.

Engines: 1.0 • 1.2 **MPG**: 38.5-47.5 **BHP**: 50-60 **Turning:** 9.2
Gearbox: Man/Auto **Boot:** 203 **Seats**: 5 **Doors:** 3/5

Nissan Micra 1992 - 2003

Proper driving school legend. Arguably the best and easiest to live with incarnation, but the older ones are getting rusty. Great little 1.3 engine. CVT gearboxes making driving a doddle except when it malfunctions. Like these 'Noddy' cars a lot. Rattly timing chains. Rusty bodywork and suspension. Mass airflow sensors and starting issues.

Engines: 1.0 • 1.4 **CO2:** 147-149g/km **MPG**: 46.3-47.1
Gearbox: Man/Auto **Boot:** 206/960 **Seats**: 5 **Doors:** 3/5

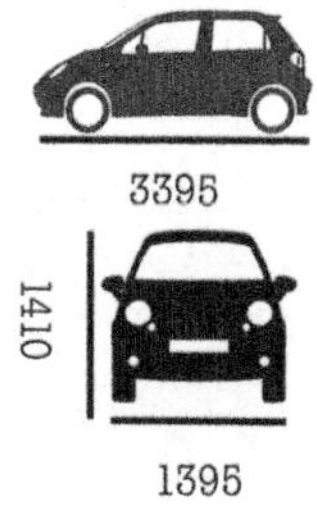

Perodua Nippa 1997 - 2002

Yet another holder of the cheapest car in Britain title. In a previous life it was a Daihatsu Mira. A basic town centre assault vehicle. Basic model was the EX and the GX delivered little more than some flash paint that was metallic and body coloured, which was no big deal. Has a great reputation for not breaking down. Body damage and rust.

Engines: 850cc **CO2:** 129g/km **MPG**: 51.9 **Gearbox:** Man/Auto **Boot:** 190 **Seats**: 4 **Doors:** 5

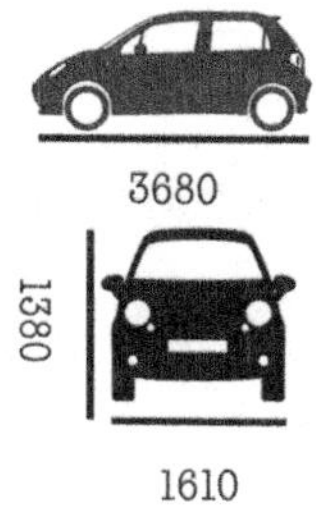

Peugeot 106 1991 - 2003

Also known as the Citroen Saxo, but with a Lion on the front. The littlest engines like the 1.0 deliver economy around town, otherwise slow, unless you go for the 1.6 GTi of course. Noisy and unsophisticated by modern standards, but who cares about modern standards? Steer clear of oil leaks and smoky diesels. Lots to like about these.

Engines: 1.1 • 1.6 • 1.5D **CO2:** 138-194 g/km **MPG**: 34.9-53.3 **Gearbox:** Man/Auto **Boot:** 215/953 **Seats**: 5 **Doors:** 3/5

Peugeot 205 1983 - 1996

One of those supermini legends without mentioning the GTi. Very light, even flimsy at times, but practical, economic and easy to get along with. Small petrol and big diesels are the ones to last a life time. Countless specials, Mardi Gras was the last of the line. Well known for oil leaks, smokey engines, noisy cams and blowing head gaskets. Still great to drive.

Engines: 1.1 • 1.4 • 1.6 • 1.9 • 1.8D **MPG**: 35.7-57.9 **Gearbox:** Man/Auto **Boot:** 216/1200 **Turning:** 11 **Seats**: 5 **Doors:** 3/5

Peugeot 206 1998 - 2009

Fun to drive and best appreciated with a missing wheel trim, Only average amount of room inside, drive before buying as it can be oddly uncomfortable for the steerer. Engines are pretty strong, overall cheap to run and economical. Perfect little Pug really. Worn lower ball joints, broken engine mounts, rattly sunroofs, ECU problems with auto gearbox.

Engines: 1.1 • 1.4 • 1.6 • 1.4D • 1.6D • 2.0D **MPG**: 45.6-65.7 **Gearbox:** Man/Auto **Boot:** 245/1130 **Seats**: 5 **Doors:** 3/5

Proton Persona Compact 1994 - 2000

Mitsubishi underpinnings is always a very good thing. Small, so cramped in the back, it is really just a runabout and not a very exciting to drive. Side impact bars, airbag and ABS on top spec models from '96. Not a lot to check for, will just be worn out, broken or not capable of getting through and MOT, if you can find one.

Engines: 1.3 • 1.5 • 1.6 **MPG**: 34.4-36.6 **Weight**: 960
Gearbox: Man/Auto **Boot:** 240/1072 **Seats**: 5 **Doors:** 3

Reliant Rialto 1981 - 1998

Replaced the Robin for a few years in order to put some distance between the recalled and troublesome old model and it came as a saloon, estate and panel van. Sat on a galvanised chassis. Styling by International Automotive Design with headlights from the Austin Metro. Some interesting '80s incarnations like the SE.

Engines: 848cc **Speed**: 80 **0-62:** 16.0 **MPG**: 43.6 **Weight**: 436 **Gearbox:** Man 4 speed **Payload:** 344kg

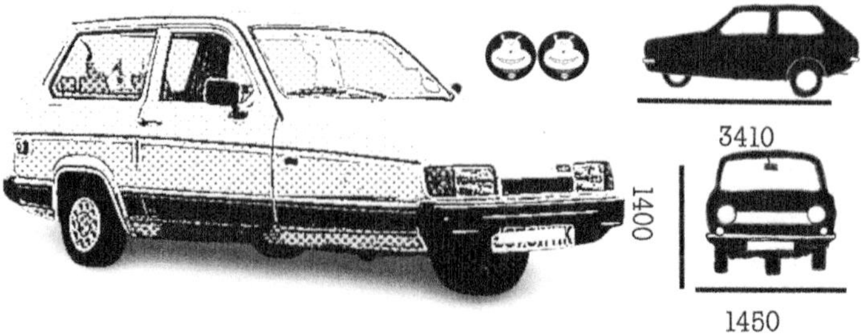

Reliant Robin Mark II 1989 - 1999

Return of the Robin with a new fiberglass body attached to a galvanised chassis, 'cool' single central windscreen wiper and a rather useful rear hatchback. Distinguished by its Fiesta frontage. There was a final 50 of these with a commemorative plaque on the dash. Basically for winter resting motorcyclists, change my mind...

Engines: 848cc **Speed**: 78 **0-60:** 13.9 **MPG**: 43.6 **Weight**: 450 **Gearbox:** Man **Boot:** 40 cu feet van

Reliant Robin Mark III 1999 - 2001

Proper face-lifted front end stretched over Vauxhall Corsa lamps, there was also remoulded doors and tailgate. There's some proper luxury touches as buyers find some deep pile carpets and uprated trim, RDS radio cassette, chrome door handles, stainless steel exhaust, fog lights and even alloy wheels. B&N Plastics version 2001-2002. Then extinction.

Engines: 848cc **Speed**: 78 **0-60:** 13.9 **MPG**:60+ **Weight**: 450 **Gearbox:** Man **Boot:** 40 cu feet van

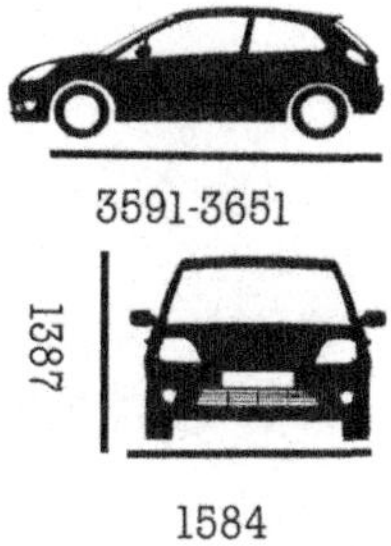

Renault 5 1984 - 1996

Yet another perfect Eurobox, this time made in France. Bags of '70s charm still, lovely to drive and inevitably cheap to run. These are late period Campus and Prima jobs, so bottom of the range fillers. 1.4 is probably the best all rounder. Build average, electrics, head gaskets, gearbox bearings, oil leaks and remember the ad 'what's yours called?' Answer: Rusty.
Engines: 1.1 • 1.4 **MPG**: 41-51.6 **Turning**: 10.25 **BHP**: 47-60 **Gearbox:** Man **Boot:** 270/900 **Seats**: 5 **Doors:** 3/5

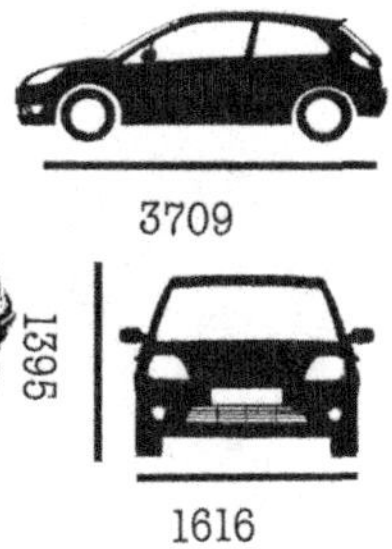

Renault Clio 1991 - 1998

Proper 'Nicole' era supercinq that was a breath of fresh air. The stylish, spacious and efficient market leader at the time. Very comfortable, the 1.4 is the best for most jobs, makes a racket, but sound proofing in '94 and bigger '96 revamp and loads of special eds. Renault build a bit average, but smoky diesels and worn out brakes, suspension etc.
Engines: 1.1 • 1.2 • 1.4 • 1.8 • 1.9D **MPG**: 32.6-50.9 **Gearbox:** Man/Auto **Boot:** 265/1055 **Weight**: 842 **Seats**: 5 **Doors:** 3/5

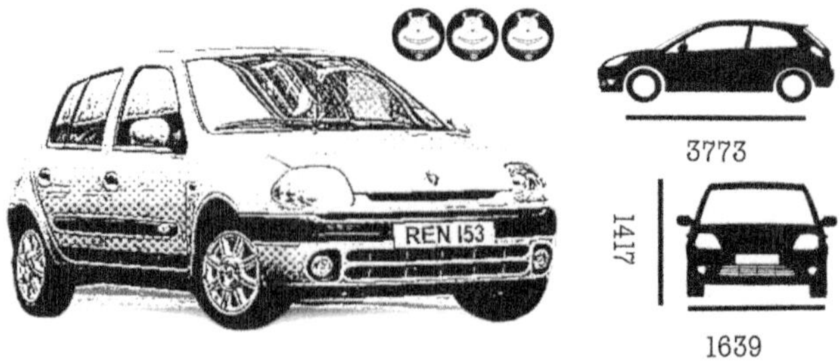

Renault Clio 1998 - 2001

This is the Mark 2 and it is bigger, so in modern terms that makes it better. Certainly felt more solid, which is good, but maybe less fun than before. Renault building their reputation for safe cars, so lots bought for first time teens. Cheap to run when not breaking down. Owners took long service intervals literally. Diesel EGR valves faulty.

Engines: 1.2 • 1.4 • 1.6 • 1.9D **CO2:** 146-179g/km **MPG**: 37.2-47.1 **Gearbox:** Man/Auto **Boot:** 255/1037 **Doors:** 3/5

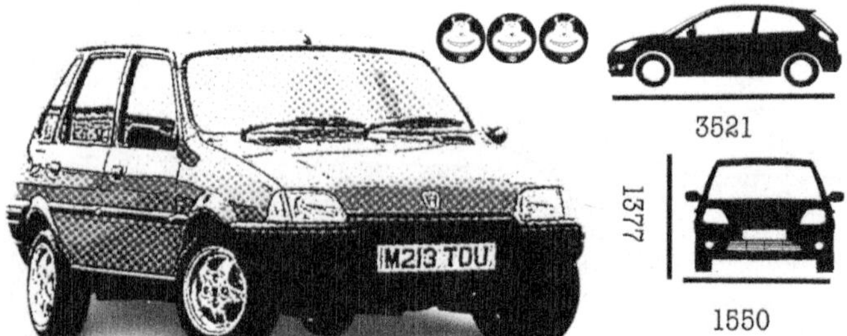

Rover Metro 1990 - 1994

Outclassed by all those brilliant French ones this remained cramped and noisy, if cheap to run and still fun to drive. Higher sped from '92 meant the L not only had a split fold rear seat, but central locking, 5 speed from '93. Plenty of issues to keep you busy beyond oil leaks, shot suspension and engine unreliability. Seems cool now. Knackered then.

Engines: 1.1 • 1.4 • 1.4D **MPG**: 38-61.7 **Gearbox:** Man/Auto **Boot:** 230/960 **Turning**: 9.8 **Seats**: 5 **Doors:** 3/5

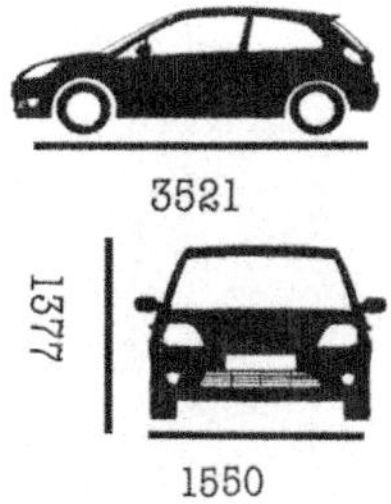

Rover 100 1994 - 1999

The very last gasp for the old thing before BMW condemned it as unsafe at any speed. Was beefed up a bit but very cramped and uncomfortable despite being a fun steer. Posh names like Knightsbridge didn't fool anyone. All the problems like oil leaks, wheel bearings, worn rear suspension and general electrical shenanigans.

Engines: 1.1 • 1.4 • 1.5D **MPG**: 40.4-56.4 **Gearbox:** Man/Auto **Boot:** 230/960 **Turning**: 9 **Seats**: 5 **Doors:** 3/5

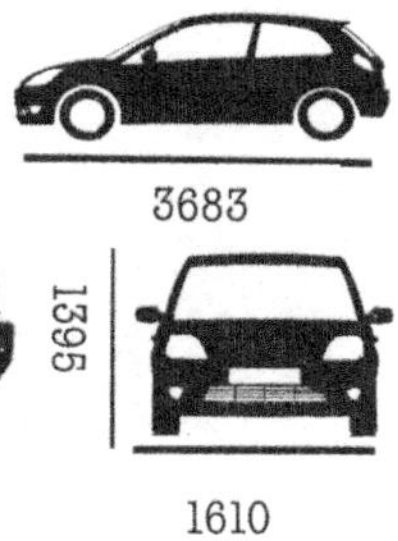

SEAT Ibiza 1985-1992

Sharp '80s styling which was inevitably by ITAL Design, the Porsche involvement was real, but mostly marketing, despite nice handling and the Fiat 903cc unit barely alive. Looks good, otherwise average supermini with a sluggish diesel and the mid range 1.2 probably the best. Oil leaks, worn front suspension and beaten panels.

Engines: 903 • 1.2 • 1.5 • 1.7D **MPG**: 39.8- 42 **Gearbox:** Man **Boot:** 320 **Weight**: 900 **Turning**: 10.1 **Seats**: 5 **Doors:** 3/5

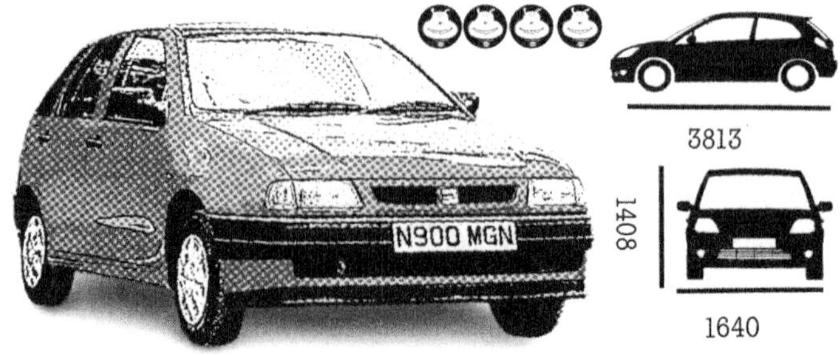

SEAT Ibiza 1993-1999

VW dig in and reinvent the Ibiza as Polo with a Spanish accent. Solid build massive choice of VAG group engines, proper badge engineering. The smaller petrols hard work, the bigger ones great and excellent diesel options. '96 revisions upped the spec and safety with seat-belt pretensioners. Clutch cable, electrics, worn suspension.
Engines: 1.0 • 1.3 • 1.4 • 1.6 • 1.8 • 2.0 • 1.9D **MPG**: 36-58.9 **Gearbox:** Man **Boot:** 335/823 **Seats**: 5 **Doors:** 3/5

SEAT Ibiza 1999-2002

The Spanish Polo. Which means it is well built, has a range of strong engines and they won't easily breakdown, unless previously owned by a cheapskate, which most are. TDI for maximum economy. Three door is on the small side at the back, but there are less common five doors. These are great. Worn suspension, timing chain tensioners, central locking.
Engines: 1.4 • 1.6 • 1.8 • 1.9D **CO2:** 138-187g/km **MPG**: 36.2-56.5 **Gearbox:** Man/Auto **Boot:** 250/782 **Seats**: 5 **Doors:** 3/5

Skoda Fabia 1999-2007

More VW magic with a Czech twist. Proper unpretentious little box and you can enjoy it even more with a 1.9 SDI which won't ever let you down. If you fancied one as a saloon, you are in luck. 2004 was a revamp for specs and bodywork. Coolant and oil leaks, timing belt tensioner, head gasket failure.

Engines: 1.2 • 1.4 • 2.0 • 1.4D • 1.9D **MPG**: 36.2-61.4 **Gearbox:** Man/Auto **Boot:** 260/1016 **Seats**: 5 **Doors:** 4/5

Subaru Vivio 1992- 1995

Early '90s SUV, in the shape of a Japanese novelty 4×4 which certainly looks cute but is pointless unless you live somewhere sticky. Not around for very long and expensive for what it was. Subarus always reliable but cost a bomb to fix even the tiddlers. Cracked cylinder heads, rust, broken transmission is the very end. Worth finding.

Engine: 658 **MPG**: 46 **Gearbox:** Man **Boot:** 106/834 **Turning**: 9.8 **Weight**: 730 **Seats**: 4 **Doors:** 5

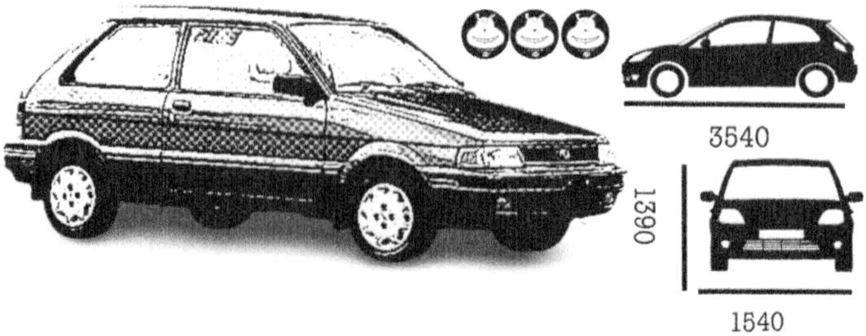

Subaru Justy 1989 - 1996

More reliable than a contemporary Range Rover on a damp Wednesday night. Certainly part time 4×4 and stepless automatic transmission blew buyers minds. Three cylinder makes a racket, decent amount of room inside, who knows the intended target market. Cost a lot new, when used it is brakes and suspension which failed first.

Engines: 997 • 1.2 **MPG**: 39.8 **Gearbox:** Man/Auto **Boot:** 198 **Weight**: 750 **Seats**: 5 **Doors:** 3/5

Subaru Justy 1996 - 2002

This is also a Suzuki Swift and built in Hungary. The Justy was more complicated as it had four wheel drive. 'Just' the one engine 1.3 and spec GX available with three or five doors. Proper oddity. No one really needs an ugly 4×4 mini hatch. Cracked cylinder heads, rust, broken transmission on 4 x 4 kills it.

Engines: 1.3 **CO2:** 165g/km **MPG**: 40.9 **Gearbox:** Man/Auto **Boot:** 250/900 **Seats**: 4 **Doors:** 3/5

Suzuki Alto 1990 - 1999

Mostly well known for being a teeny tiny automatic combined with a buzzy three cylinder engine. Slow but it delivered plenty of mpg. Quite dated and pointless at the time and it seemed to disappear of the spec sheets and out of the showrooms as there was so many other Justy and Swift Tiddlers to choose from.
Engines: 796 **MPG**: 50.9 **BHP**: 40 **Gearbox:** Man/Auto **Boot:** 210/1200 **Weight**: 705 **Seats**: 4 **Doors:** 3/5

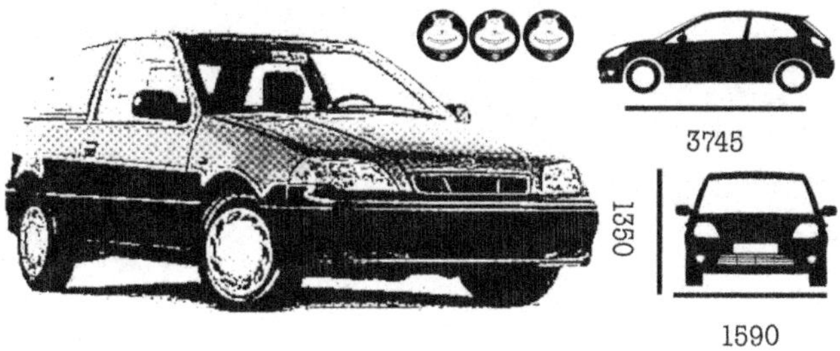

Suzuki Swift 1992 - 2003

Proper back to basics Japanese boxes with nothing going for them apart from reliability and fuel economy. What more do you want? Split fold rear seats standard. 1.0 and 1.3 engines plus a fun filled GTi. Running costs are reasonable enough. all dead. So simply wear and tear and how many have survived abuse/neglect? **Engines:** 1.0 • 1.3 **CO2:** 130g/km **MPG**: 51.4 **Gearbox:** Man/Auto **Boot:** 249/579

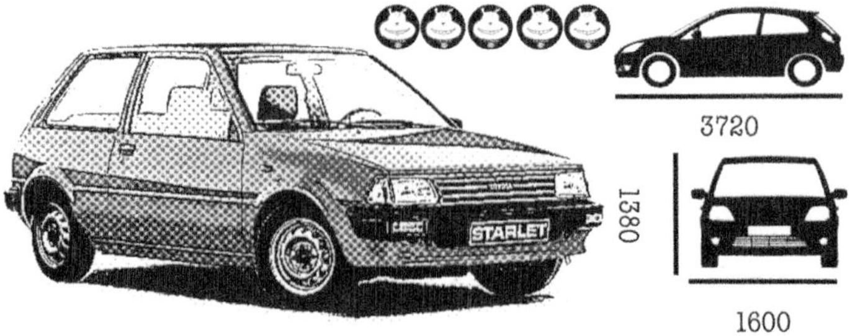

Toyota Starlet 1990 - 1996

Not rated by 'experts' at the time for its lack of character and overall dullness, except that buyers actually appreciated their utter reliability. Got side impact bars in 1992. Electric sunroof and split folding seats originally. But, UK fitted sunroofs often leaked, so most will have rusted to its constituent parts.

Engines: 1.0 • 1.3 **MPG**: 37.1-42.1 **Gearbox:** Man **Turning**: 4.6 **Boot:** 226/490 **Weight:** 805 **Seats**: 5 **Doors:** 3/5

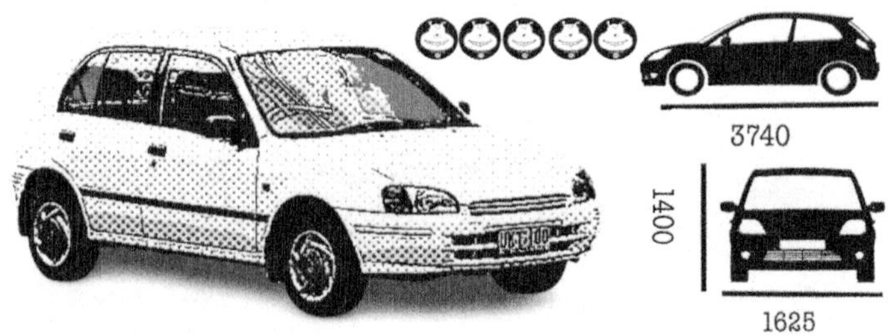

Toyota Starlet 1996 - 1999

Better packaged and priced than before but just hanging around until the Yaris arrived, indeed the two models overlapped. CD and GLS have decent equipment levels from may '98 there's a larger airbag and uprated suspension. Special editions are mostly just paintwork. Tough of course, just minor electrics and abused clutches.

Engines: 1.3 **MPG**: 42.2 **Gearbox:** Man/Auto **Boot:** 226/490 **Weight:** 835 **Seats**: 5 **Doors:** 3/5

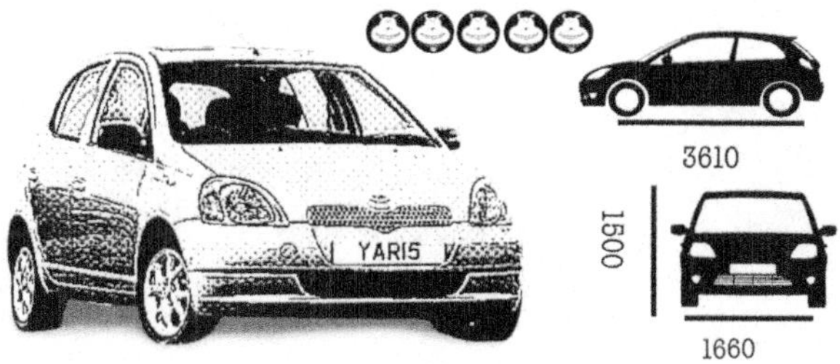

Toyota Yaris 1999 - 2006

Here's the no brainer buy with 1.0 and 1.3 engines, model was revamped in 2003. Looks like a tiny conventional hatch from the outside, but actually very roomy inside and a ton of cubbyholes. Decent equipment levels from the S upwards. Automatic gearbox can fail, early model head gaskets. Overall takes a lot of abuse.

Engines: 1.0 • 1.3 • 1.5 • 1.4D **CO2:** 117-134g/km **MPG**: 50.4-64.2 **Gearbox:** Man/Auto **Boot:** 205/950 **Seats**: 5 **Doors:** 3/5

Vauxhall Nova 1983 - 1993

Boxy basic hatch which was a people's favourite being practical, easy to fix, despite the noise and overall coarseness. Getting old in '90s. Luxe models had velour and 1.4s five speed boxes, late Luxe+ electric windows and central locking. Oil leaks, noisy cams and worn gearbox linkage amongst other high mile/use ailments.

Engines: 1.0 • 1.2 • 1.4 • 1.6 • 1.5D **MPG**: 38.3-56.1 **Gearbox:** Man **Boot:** 225 **Turning**: 9.4 **Seats**: 5 **Doors:** 3/5

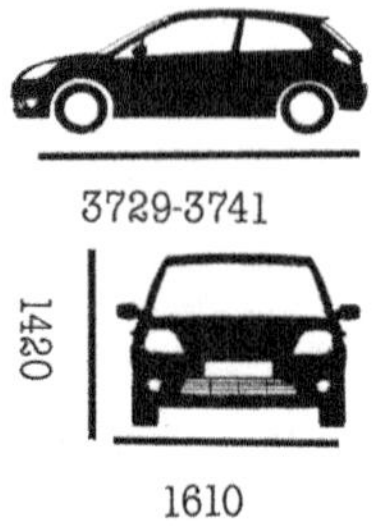

Vauxhall Corsa 1993 - 2000

Basic supermini motoring with 1.0 and 1.2 petrols are crudely adequate. 1.4 better if the commute is more dual carriageway. Decent economy from the small petrols and maximum mpg from the diesel of course. Another driving school favourite. Avoid those ones. Steering racks, wheel bearings, water pumps and timing belt servicing.

Engines: 1.0 • 1.2 • 1.4 • 1.6 • 1.5D • 1.7D **MPG**: 48.7-62.8 **Gearbox:** Man/Auto **Boot:** 260/1060 **Seats**: 5 **Doors:** 3/5

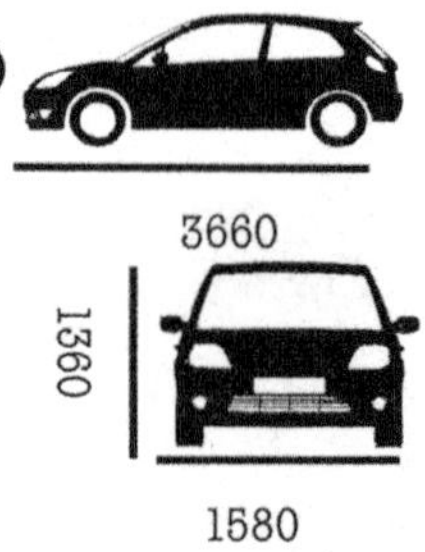

Volkswagen Polo 1990 - 1994

Brilliant Bread Van original face lifted for the '90s. More of the same but much more solider, there's still a rather pointless saloon. Servo brakes are better, but everything is an extra, just remote door mirrors on the CL. Pretty basic and bullet proof, blown head gaskets, broken starters, leaking brake cylinders. Remains underrated.

Engines: 1.05 • 1.3 **MPG**: 37.9-39.6 **Gearbox:** Man **Weight**: 767 **Boot:** 310/1224

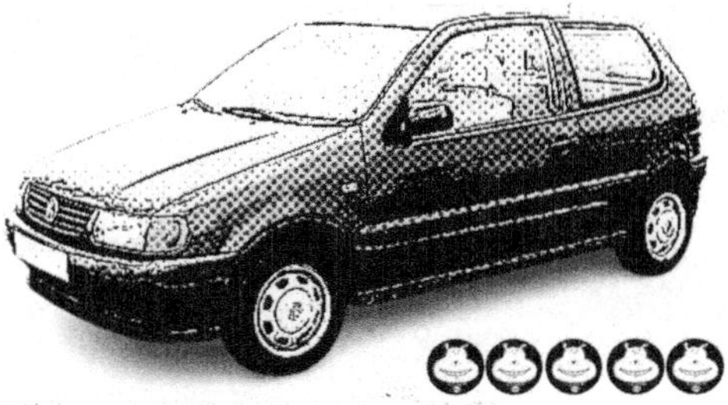

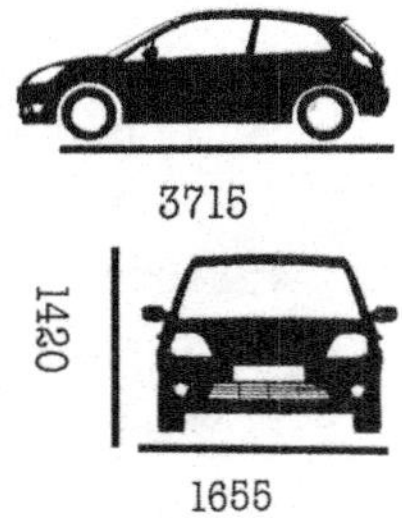

Volkswagen Polo 1994- 1999

Peak '90s Polo which is sophisticated, comfortable and quite refined. Saloon is still about, search around the top specs for ABS and airbags. CL and GLX has most of what you need. The tiny petrols do struggle, the 1.4 is the best all rounder and the 1.9 diesel slugs on and on for maximum mpg. Oil leaks, clutch release cable a bit sticky. Mostly owner neglect. **Engines:** 1.0 • 1.3 • 1.4 • 1.6D • 1.9D **MPG**:36.2-49.3 **Gearbox:** Man/Auto **Boot:** 455/760 **Circle**: 10.6 **Seats**: 5 **Doors:** 3/5

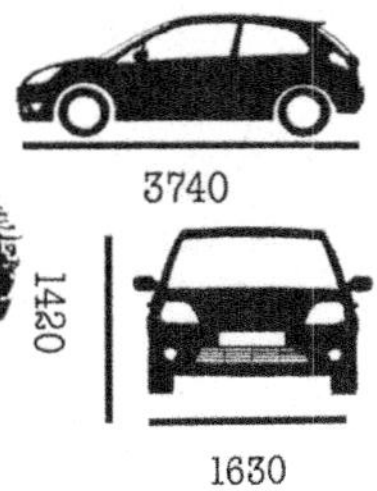

Volkswagen Polo 1999 - 2001

Beefed up Polo re-engineered to be a surrogate Golf making it quite a sharp drive. Incredibly there is still a saloon to enjoy. Lots of engines to choose from and big car features. On the whole these are tough and reliable runabouts and get rarer by the day. Difficult to kill. Can chew through tyres and brakes, mostly wear items to look at and rust. Gearbox failure. **Engines:** 1.0 • 1.4 • 1.6 • 1.4D • 1.9D **MPG**: 40.4-62.8 **Gearbox:** Man/Auto **Boot:** 245/975

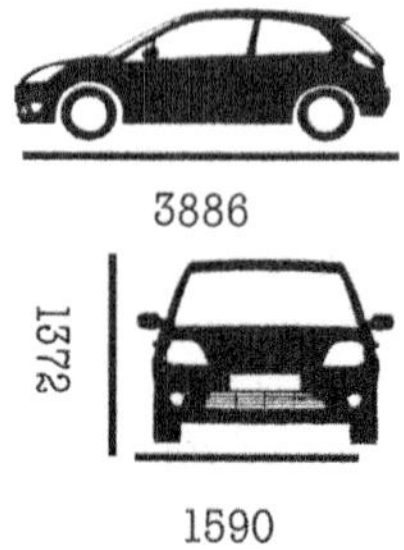

Zastava Yugo 311 • 511 • 513 1971 - 1990

In a previous life this was a Fiat 128 saloon. Ancient tech with coarse mechanicals of course. Rarely cherished, mostly endured because it was cheap and cheerless. Not a great deal to say about these, although technically in circulation for decades, now extinct. Being rough and basic has its advantages, smoke, oil leaks and rust are a given.

Engines: 1.1 • 1.3 **MPG:** 38.2 **Gearbox:** Man **Boot:** 325/1010 **Seats**: 5 **Doors:** 3/5

Family Bangers - Commute Shop Repeat

Eventually I'll have to call this the Golf Class, but in the '90s Volkswagen had a bit of a wobble. They recovered, but so many other manufacturers caught up with what VW were up to. Indeed, the '90s is special because in came the game changing Ford Focus. Great to drive, cutting 'edge' looking, but most importantly for Bangerpedia purposes, very excellent in day to day service. Reliable, easy and cheap enough to fix and just gets on with the job. The Escort was outdated and outclassed, but it never went away as there were hardcore loyalists who appreciated its old fashioned grimness.

The Japanese contribution was spectacular as usual and the main players models really are proper mid vehicle sized legends that begin with the letter C. Corolla and Civic, very good in the '80s, these became mainstream ultra reliables which got real families through the hard times. Honestly, the direct real world experience I have and the feedback from users and owners is that these were the very best Bangers ever made.

There are also lots of overlooked Japanese models that are just as impressive as the big Cs. The Mazda 323 came into its own and the middle order Nissans never gave up, the Sunny always shone and the ultra dull Almera would keep on going without any complaint. The Mitsubishi Lancer was a decent collection of parts which served their owners so well.

Possibly the best thing about going back to the '90s is that some very old friends start turning up, so the positively ancient Renault 4 makes a welcome guest appearance.

Oh yes, and this is where Rover finally became good, with considerable assistance from Honda, but the 200 is now a used car worth buying. The upmarketisation of this section was complete with the Audi A3, which of course was just a posher Golf and also the interesting reworking of the old 3 Series into the Compact.

In many ways this is the most competitive and sometimes confusing sector where almost anything goes.

Alfa Romeo 33 1983 - 1995

Grown up Alfasud, revamped in '89 and their biggest ever seller. Now had a galvanised body, revised rear suspension, all had power steering and fuel injection. Good specifications except basic 1.5 had no rear wiper. Sportiva and Turismo, had fragile Alcantara trim and comfy Recaros. Engines need looking after. **Engines:** 1.3•1.5•1.7. **MPG**: 35.8-36.7 **Gearbox:** Man **Boot:** 350 **Weight**: 910 **Seats**: 5 **Doors:** 5

Alfa Romeo 145-146 1994 - 2000

Alfa's Golf and could have been the next Lancia Delta. Better handling from 1996, along with the Twin Sparks. L had electric windows and TS rather more, including power steering. All the early flat fours should have belt and tensioner changes when bought. Electricals not that bad, warning light for ABS and engine that stay on...

Engines: 1.6•1.7•1.8•2.0 **MPG**: 31-34.7 **Gearbox:** Man **Boot:** 320 and 380 **Seats**: 5 **Doors:** 3/5

Audi A3 1996 - 2003

Rebirth of the posh hatchback (Audi 50), a Golf with knobs on, not forgetting the three door Sportback. If you want to get places though there is a 2.0 turbo, although the 3.2 V6 is sensationally quick. Five doors from '99, never as much fun as a BMW 3 series. Diesel 1.9 just a Golf TD. Air flow sensors, rusty brake lines, engine management issues.

Engines: 1.6 • 1.8 • 1.9D **CO2:** 135-226g/km **MPG**: 30.1-57.7
Gearbox: Man/Auto **Boot:** 350/1100 **Seats**: 5 **Doors:** 3/5

BMW 3-Series Compact 1994 - 2001

Frankenstein hatchback part '90s 3 (E36) and '80s E30, almost like BMW forgot they made the Touring. It's really a 2 plus a very occasional two. Good old fashioned fun though. Small petrol and diesel engines suit it. Problems include sensors, blown head gaskets so best to buy new water pump as a precaution. Suspension and all sundries.

Engines: 1.8 • 1.9 • 1.9D **CO2:** 183-211g/km **MPG**: 37.2-47.9
Gearbox: Man/Auto **Boot:** 306/1030 **Seats**: 5 **Doors:** 3

Chrysler Neon 1999 - 2003

Misunderstood from day one, even when revamped. SE a low spec pointless exercise, better go for the leather lined LX if you really must have a 2.0 litre hatch with a decent amount of space for people and their luggage. A curiously dated American. Head gaskets can go, air con failure, tired suspension parts.

Engines: 2.0 **CO2:** 185-196g/km **MPG**: 25.2-35.3 **Gearbox:** Man/Auto **Boot:** 371 **Seats**: 5 **Doors:** 4

Citroen ZX 1991 - 1996

Less sexy version of the Pug 306 in various formats. gets on with family job and with the unburstable basic diesels that will run on chip fat make it a cockroach used buy. Revisions in '94, restyle and seatbelt pretensioners. Electrics can and will play up. Worn out suspension, starting and running problems due to blocked injection systems.

Engines: 1.4 • 1.6 • 1.8 • 2.0 • 1.9D **MPG**: 33.3-49.1 **Gearbox:** Man/Auto **Boot:** 324/1146 **Seats**: 5 **Doors:** 3/5

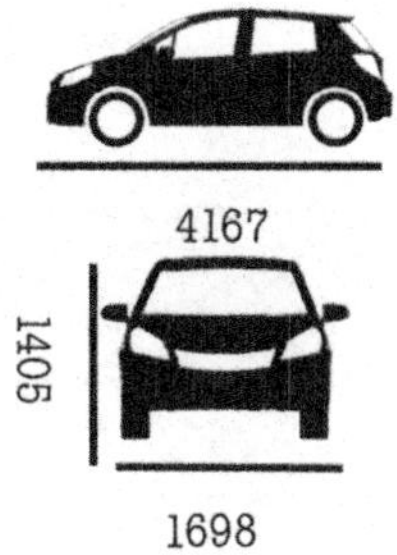

Citroen Xsara 1997 - 2000

Proper old school basic Frenchie. A sexed up ZX which was always reliable enough and pretty frugal on the old fuel whether you go for a 1.4 or 2.0 HDi. Does not seem to have many reliability downsides provided it has not been run into the ground. ECU fails, some late '90s diesel engine issues can kill the engine. Diesel clutches, also tatty bodywork.

Engines: 1.4 • 1.6 • 2.0 • 1.9D • 2.0D **MPG**: 40.9-54.3
Gearbox: Man/Auto **Boot:** 408/1190 **Seats**: 5 **Doors:** 5

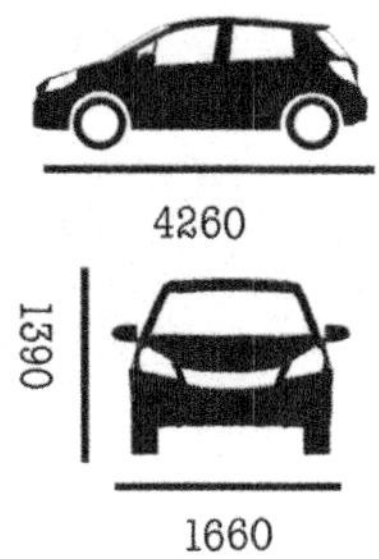

Daewoo Nexia 1995 - 1997

An '80s Astra/Belmont, no bad thing, came with standard power steering, ABS, airbag and the Isuzu engines are reliable GL is 8v and GLi 16v. GLXi has all the spec including electric sunroof, central locking, split rear hatch seats. These get driven into the ground, so it is all the wear and tear parts. Exhaust, brakes etc. **Engines:** 1.5 **MPG**: 38-42.8
Gearbox: Man/Auto **Boot:** 390/530 **Seats**: 5 **Doors:** 4/5

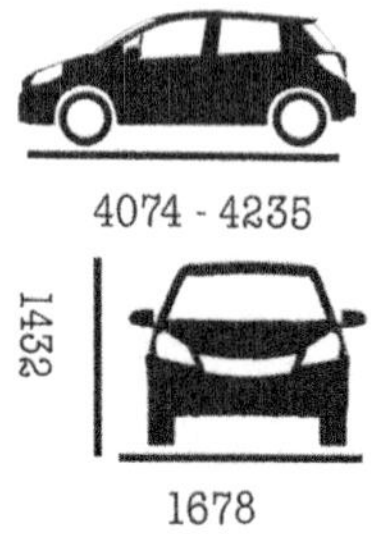

Daewoo Lanos 1997 - 2002

Hard to be enthusiastic about these blobby hatches and saloons. They do exist and largely don't break down, but they suffer from neglect because no one cares so long as they work. Parts can cost more than you would ever believe. A last resort purchase for many. Underrated. Not much goes wrong, just old age issues like brakes and suspension.

Engines: 1.4 • 1.6 **MPG**: 34.4-36.6 **Gearbox:** Man/Auto **Boot:** 320/960 **Seats**: 5 **Doors:** 3/4/5

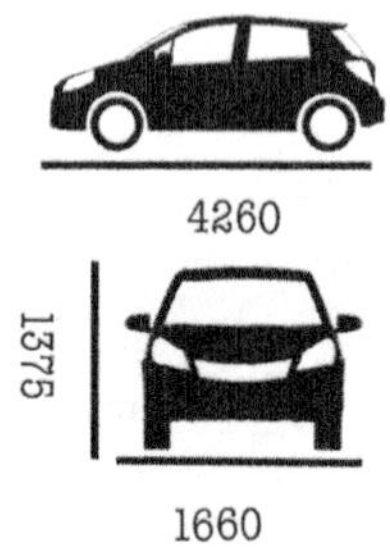

Daihatsu Applause 1990 - 1995

Interesting oddity that never caught on. A hatchback that looked like a saloon wasn't something that the world, or at least UK market wanted. Sluggish carb, slightly less sluggish injected engine. Has power steering GLXi alloys and two tone paintwork. OAP special also had an afterlife as a minicab. Parts now a problem.

Engines: 1.6 **MPG**: 39.7 **Gearbox:** Man **Weight**: 930 **Turning**: 9.6 **Boot:** 412/764 **Seats**: 5 **Doors:** 5

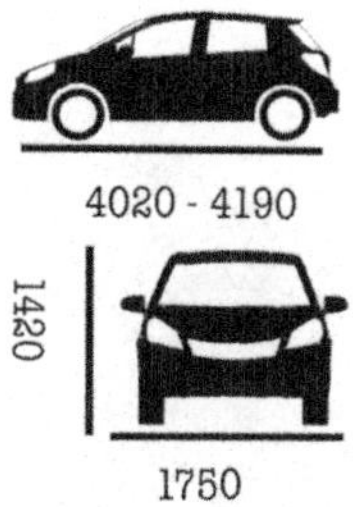

Fiat Brava/Bravo 1995 - 2002

Your choice, a three door or a five door with different names and some neat design twists (trick wheel jack). Still a Fiat though, doesn't make it bad, but means they have been neglected despite now being rarer than most contemporary Ferraris. Buy 1.6 petrol and TD100SX. Worn suspension, catalytic converter, blocked radiators, electric windows.

Engines: 1.2 • 1.6 • 1.9D **CO2:** 146-186g/km **MPG**: 36.2-51.2
Gearbox: Man/Auto **Boot:** 379/1164 **Seats**: 5 **Doors:** 3/5

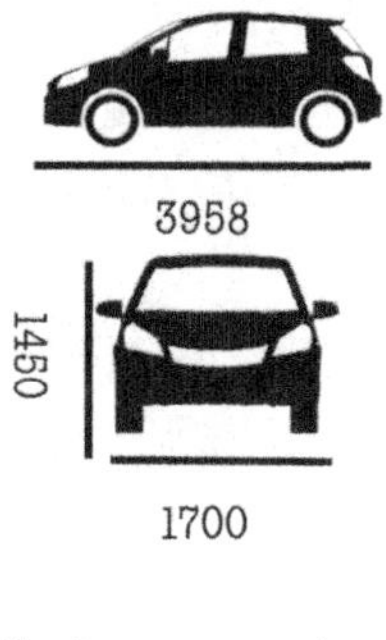

Fiat Tipo 1988 - 1995

One of the best of the '80s alt-Golf hatches that was spacious and fun to drive which lived on in Turkey till 2000. upgraded in '93: side impact beams and airbags, in '94. Some find ride harsh, but it is massively practical. Just the usual Fiat flimsiness, so electrical issues, smokey diesels, oil and coolant leaks. Cambelt change if and when you can buy one.

Engines: 1.4 • 1.6 • 1.8 • 2.0 • 1.7D • 1.9D **MPG**: 29.3-46.1
Gearbox: Man/CVT **Boot:** 350/710 **Seats**: 5 **Doors:** 3/5

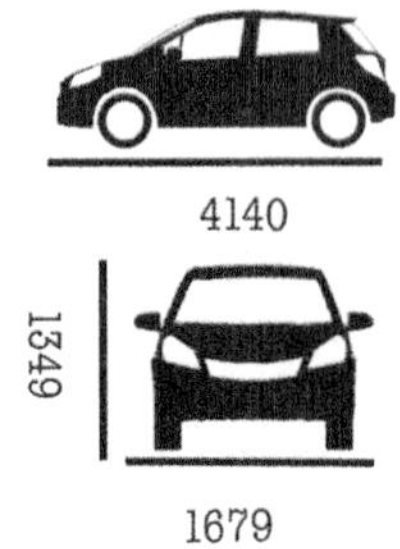

Ford Escort/Orion 1990 - 1997

Regarded as a cynical collection of new parts that produced a dreary and underwhelming old thing. Still sold in big numbers, got revamped in '92 with restyle and new Zeta engines. 4 door Orion, badged as saloon from '93. Electricals never great, ECUs and windows, noisy camshaft, tappets, snapped cam belts. **Engines:** 1.3 • 1.4 • 1.6 • 1.8 • 1.8D **MPG**: 25.6-53 **Gearbox:** Man/Auto **Boot:** 379/1130 **Seats**: 5 **Doors:** 3/4/5

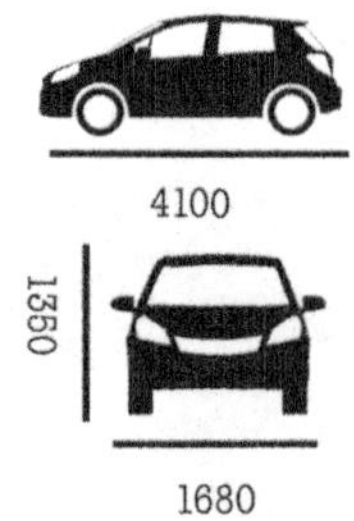

Ford Escort 1995 - 2001

Mark 7 lived on as a special edition in the Focus era. Flight and Finesse with 1.6 petrol and 1.8 Turbo Diesel. Rust, neglect and abuse killed most of them. Looks and drives prehistoric, but a solid used buy. Blown head gaskets, bodywork rust serious, rusty fusebox, can be fatal. **Engines:** 1.6 • 1.8D **CO2:** 167-175g/km **MPG**: 42.7-44.2 **Gearbox:** Man/Auto **Boot:** 380/1130 **Seats**: 5 **Doors:** 5

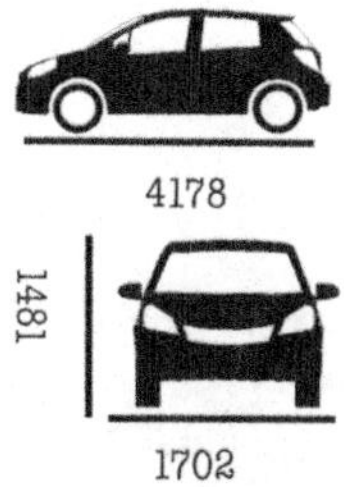

Ford Focus 1998 - 2004

Everyone loves a Focus and they are great to drive and easy to own. Hard to go wrong with any of them, groovy looking three door still a thing, oh and briefly a saloon. TDCI diesels are great Zetec spec and the petrols are all pretty reliable. Getting old and scrappy and starting to rust. Otherwise it is just wear and tear to the suspension. **Engines:** 1.4 • 1.6 • 1.8 • 2.0 • 1.8D **CO2:** 140-188g/km **MPG**: 30.0-53.3 **Gearbox:** Man/Auto **Boot:** 350/1205 **Seats**: 5 **Doors:** 3/5

FSO Polonez 1978 - 1991

How to make the original utterly unexceptional Fiat rebadged product even worser, just stick a big hatch body on top. Had a radio cassette. Still terrible to drive. Unpleasantly distracting especially when re-designated as LE with fiberglass extensions. Engine went from 1.5 to 1.6 in 1989. The Poles made over a million. State sponsored sales...

Engines: 1.6 **MPG**:40 **Gearbox:** Man **Weight**: 1274 **Boot:** 425/800 **Seats**: 5 **Doors:** 5

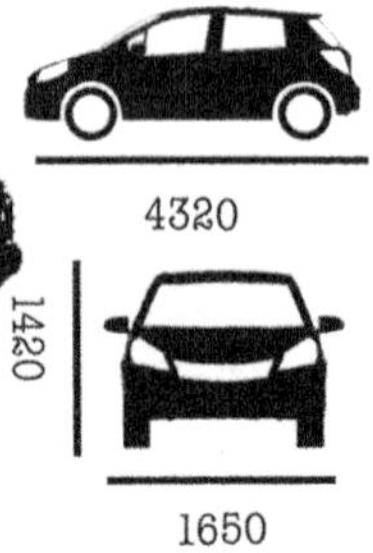

FSO Caro 1993 - 1998

Just when you thought FSO had been crushed by the fall of the Iron Curtain, they struck a deal with Rover, for the 1.41 K-Series petrol engine, we did get the solid Citroen 1.9 diesel. They had a saloon and a truck version. In 1997, Daewoo took over FSO and built their stuff. The Caro limped on to 2001 when Daewoo went broke.

Engines: 1.6 • 1.9D **MPG**: 42mpg **Gearbox:** Man **Weight**: 1274 **Boot:** 425/800 **Seats**: 5 **Doors:** 5

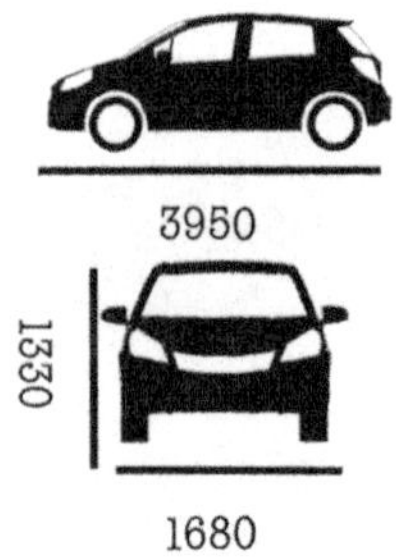

Honda Civic 1987 - 1991

Great little cars. Which did come as four wheel drive and estates, but not always available in the UK. Great for shopping and could be spiced up to deliver lots of fun. Mechanicals will go on forever, just more difficult and more expensive to fix than Escort. Nothing looks cooler.

Engines: 1.3 • 1.6 **MPG**: 38.2-42.8 **Gearbox:** Man **Weight**: 835-1025 **Boot:** 250 **Seats**: 5 **Doors:** 3/5

Honda Civic 1991 - 1995

More Civic greatness which were built to exceptional standards. Hatches and saloons are the staple models, disc brakes all round on 1.6s. Three door is essentially coupe shaped, lots of special editions, basic DX, otherwise LS and ES models with rear wings and things. Very durable and noisy cams plus smoke means its never been cared for.

Engines: 1.3 • 1.5 • 1.6 **MPG**: 33.5-47.5 **Gearbox:** Man/Auto
Boot: 196-337 **Weight:** 850 **Seats**: 5 **Doors:** 3/4

Honda Civic 1995 - 2001

Blueprint Banger. Really, owners will tell you just how brilliant these are. It is also a Rover 400, but better. Great petrol engines and comfortable ride. Everything else seems average from space to performance, but it is just relentlessly capable. That's a good thing. No real issues all, rust is probably more of an issue which can kill it.

Engines: 1.4 • 1.5 • 1.8 **MPG**: 32.1-47.5 **Gearbox:** Man/Auto
Boot: 225/600 **Seats**: 5 **Doors:** 3/5

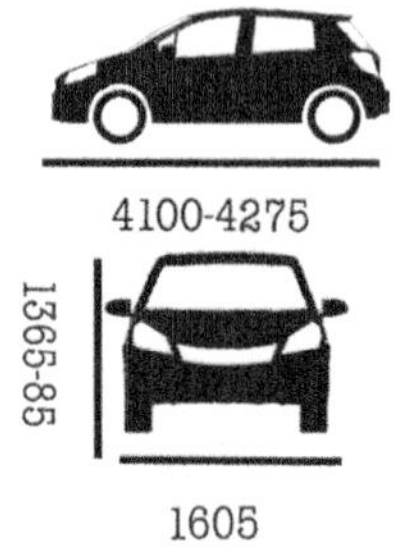

Hyundai Pony X2 1990 - 1994

Nothing new, just a revamp of the mid-80s Giugiaro-styled Pony which looks like a pint sized Sonata. Licence built Mitsubishi engines are ultra-reliable, otherwise unremarkable to drive and quite dated. Decent equipment levels of course and it would last forever whether you wanted it to or not.

Engines: 1.3 • 1.5 **MPG**:37.7 **Gearbox:** Man/Auto **Boot:** 325/415 **Turning**:9.8 **Seats**: 5 **Doors:** 3/4/5

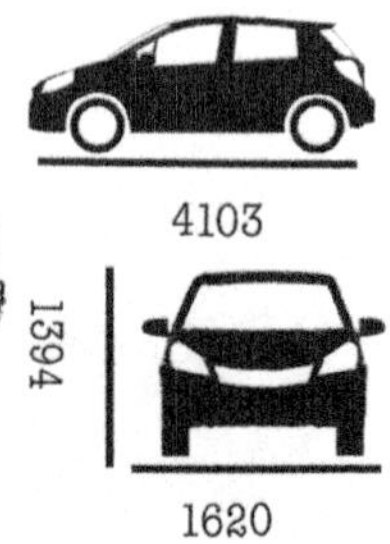

Hyundai Accent 1994 - 2000

Unremarkable old school Hyundai. Cheap and cheerful overall with tacky trim, but fine on the road where it counts. Top spec GLSi has split/fold boot/tailgate. Noisy engines not overly economical, but they are reliable. Body dents easily, can have worn suspension, leaking power steering, uneven tyre wear and dodgy heater.

Engines: 1.3 • 1.5 **MPG**: 33.2-38.2 **Gearbox:** Man/Auto **Boot:** 320 **Seats**: 5 **Doors:** 3/4/5

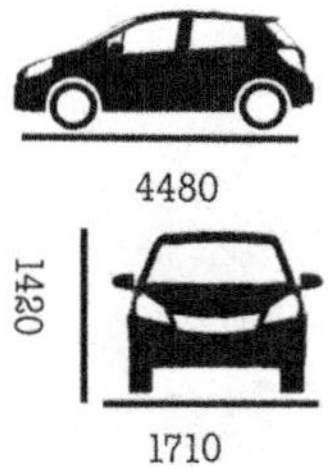

Kia Shuma 1999 - 2001

This is a hatchback Mentor I'm afraid. There was a 1.5 and 1.8 and it had had some extras but it is just a hatch. When new it was cheap, when used it has to be really, really cheap as there is nowt else apart from reliability to recommend it. Might be rusty and need a brake overhaul and service. Its unremarkability means it deserves 5 Slogs but it only gets...

Engines: 1.5 • 1.8 **CO2:** 185g/km **MPG**: 31.0-35.8 **Gearbox:** Man/Auto **Boot:** 328 **Seats**: 5 **Doors:** 5

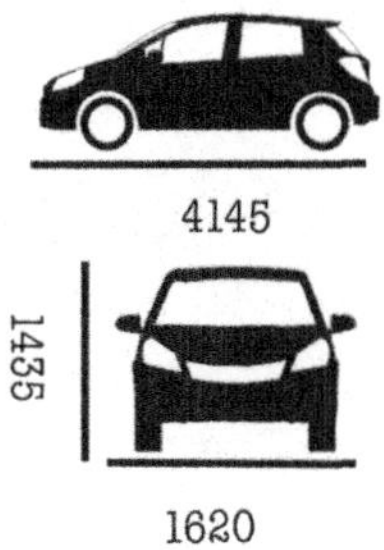

Lada Riva 1985 - 1996

Cheap and cheerless old Fiat that was shockingly popular on account of its dirtcheapness. Not much to recommend it, especially a few neglectful owners down the line. Range revised in 1991 into Es and Ls, '92 revamp as 1.5L gets colour coded grilles, remote adjustable door mirrors and a stereo cassette. Cats from '93. Noisy diff, knocking big ends.

Engines: 1.3 • 1.5 • 1.6 **MPG**: 29.7-31.5 **Gearbox:** Man
Boot: 385 **Weight**: 1057 **Seats**: 5 **Doors:** 4

Lada Samara 1988 - 1996

This is properly Russian so it is pretty dreadful. Ugly, coarse and shoddy. Terrible new, much worse, used. Has lots of low level easily added, insignificant kit like radio cassettes, sunroofs and coach lines. Rebadged in 1994 with colour coding to make it look like a real car. Usually on its last remoulds when up for sale. Don't bother.

Engines: 1.1 • 1.3 • 1.5 **MPG**: 36.1-39.8 **Gearbox:** Man
Boot: 330/940 **Weight**: 930 **Seats**: 5 **Doors:** 3/4/5

Lancia Delta 1979 - 1994

Boxy '70s hatch which is best known for being a rallying winning blistered arched legend. It also was fine at shopping., but indifferently built and very dated by the '90s. Averagely capable 1300, whilst 1600 has a bit more zing. Hard to find a cooking model anymore which will be rusty, with smokey tired mechanicals and collapsing trim.

Engines: 1.3 • 1.6 • 2.0 **MPG**: 35.3 **Gearbox:** Man/Auto **Boot:** 260/940 **Seats**: 5 **Doors:** 5

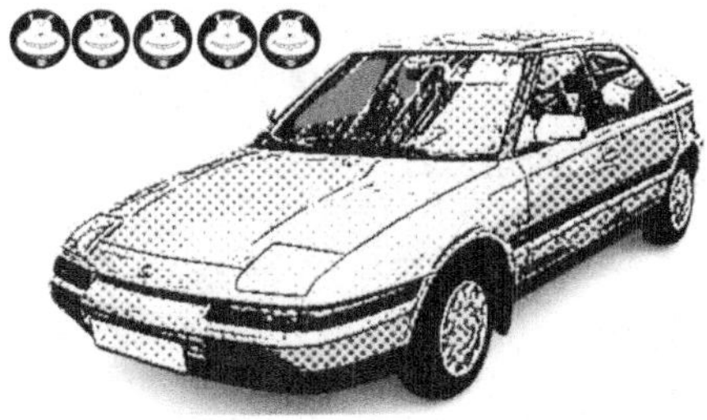

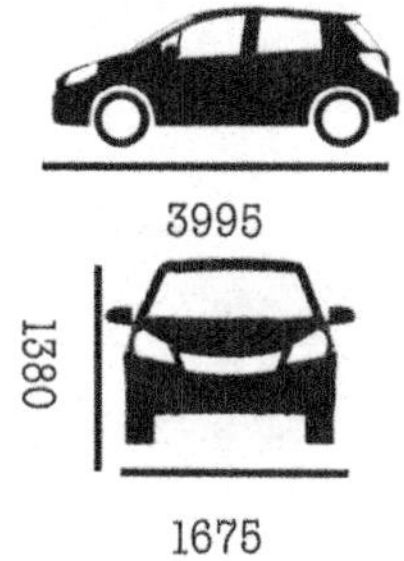

Mazda 323 1989 - 1994

There are several styles and its the 5 door fastback which we all know and love especially with pop up headlamps. Injected engines from 1991, modded suspension and an optional automatic with the 1.6 engine. Side impact beams from '93. Overall incredibly reliable and hard to fault unless neglected + servicing skipped.

Engines: 1.3 • 1.6 • 1.8 **MPG**: 33.8-38.3 **Gearbox:** Man/Auto **Boot:** 309 **Seats**: 5 **Doors:** 3/4/5

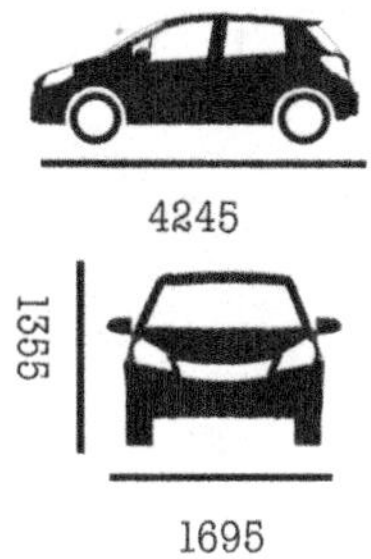

Mazda 323 1994 - 1998

Still going strong in all formats, but the Fastback gets all the attention. V6 2.0 rather exciting, basically a 5 door MX-5 in handling department. All basic LXi had twin airbags. GLX has alloys and heated mirrors, loads of pointless special editions. The whole point of this generation 323 is that it remains staggeringly reliable. Expensive to fix though.

Engines: 1.3 • 1.5 • 1.8 • 2.0 **MPG**: 29.1-38.2 **Gearbox:** Man/Auto **Boot:** 416 **Seats**: 5 **Doors:** 3/4/5

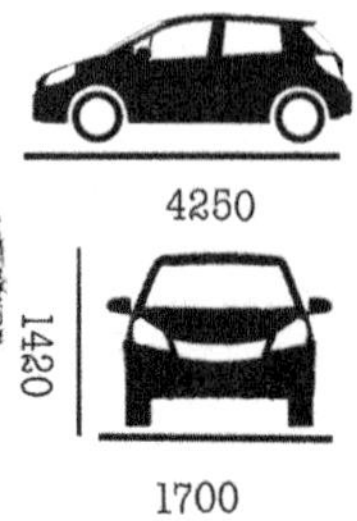

Mazda 323 1998 - 2003

These were great. Even if they didn't have the pop up headlamps anymore. Great to drive and tough as absolute nuts. There are the tiny petrols which will run all day. 1.8 and 2.0 are pokey and great of you intend on ambitious journeys. These go on forever, getting quite old and tired, will need shock absorbers, broken electrical switches.

Engines: 1.3 • 1.6 • 1.8 • 2.0 **CO2:** 179-203g/km **MPG**: 33.2-38.2 **Gearbox:** Man/Auto **Boot:** 356/697 **Seats**: 5 **Doors:** 5

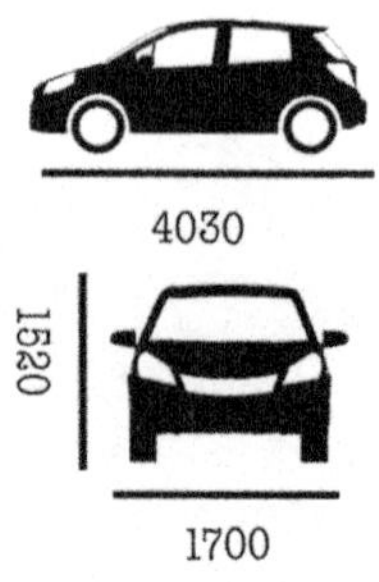

Mitsubishi Space Star 1999 - 2005

You could put this in the People Shifter category but that would be daft. It is just a big hatchback Colt which makes it a small family car. Rear seats are not removable, so smaller and less useful than a Scenic. Very reliable by virtue of being a Mitsubishi which compensates for a lot. Gasoline Direct engine is great. Wear and tear is the biggest enemy.

Engines: 1.3 • 1.8 **MPG**: 40.4-41.5 **Gearbox:** Man
Weight: 1075 **Boot:** 370/1370 **Seats**: 5 **Doors:** 5

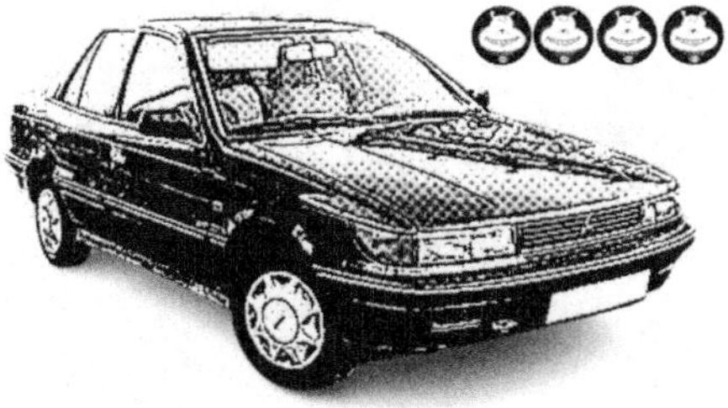

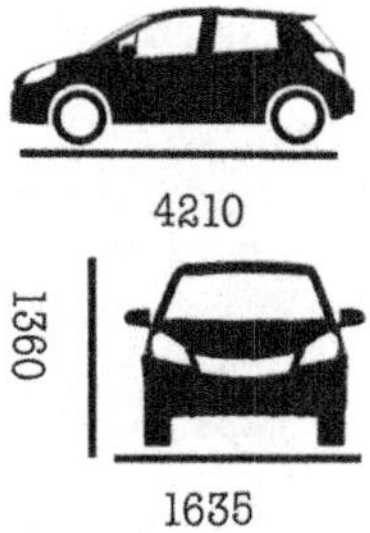

Mitsubishi Lancer 1987 - 1991

Not a great deal to write home about, just a very reliable and dull collection of saloons and hatches. Longer wheelbase saloons, so more rear legroom. Top of the range GTis and also four wheel drive versions. Hard to kill but the nation still bought cheap to fix Escorts. Ones that are left will simply be overtired and need exhausts + alternators.

Engines: 1.3 • 1.5 **MPG**: 32.7-38.2 **Gearbox:** Man **Boot:** 337 **Weight**: 975 **Seats**: 5 **Doors:** 5

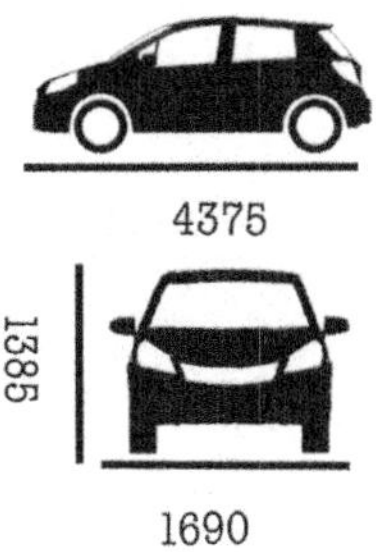

Mitsubishi Lancer 1992 - 1995

Just hatches for now. Not a lot of spec, but owners get split rear seats, tinted glass and a four speaker stereo and it was possible to get an optional auto box with the 1.6 engine. Tough as nuts of course and often became a low rent minicab with the reasonably economical 1.2. Mirage was a special edition, but lots sold with fake wood and colour coding. Nice.

Engines: 1.3 • 1.6 • 1.8 **MPG**: 35.1-40.0 **Gearbox:** Man/Auto **Boot:** 470 **Turning**: 9.6**Seats**: 3/5 **Doors:** 5

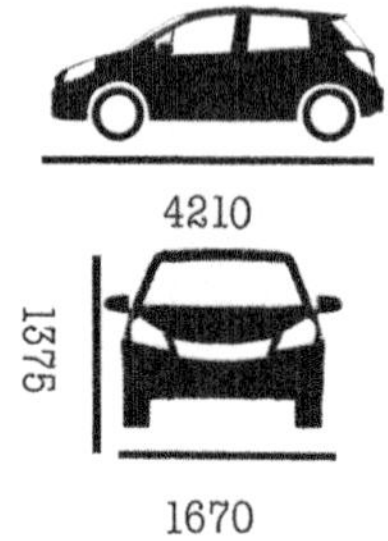

Nissan Sunny 1991- 1995

Legendary old lag that made it into the '90s in good health. Always great value and although the GT had fans it was mostly cashed strapped families and OAPs that invested heavily in this dull but worthy wagon. '93 saw injection engines and side impact beams through the range. Old age means oil leaks, chirping camshaft and collapsing trim.

Engines: 1.4 • 1.6 • 2.0 • 2.0D **MPG**: 26.1-47.4 **Gearbox:** Man/ Auto **Boot:** 490 **Seats**: 3/4/5 **Doors:** 5

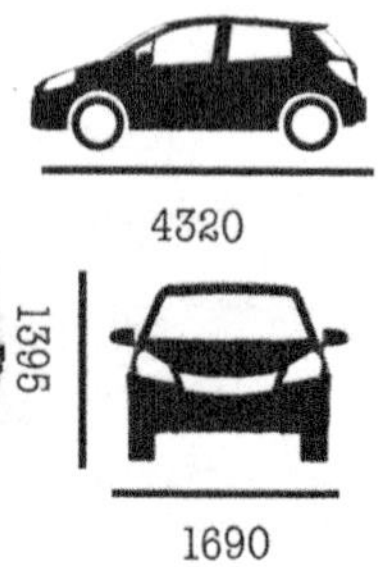

Nissan Almera 1995- 2000

Practical, certainly not stylish, but solid, safe and boringly reliable. Not bad equipment levels, saloons all have electric mirrors and radio cassette, Si for instance had a snazzy CD and sports seats. Air con from '97 on SLX, SRi and GTi and an unbelievable number of special editions for the '99 model run out. Late life wear and tear to suspension and engine.

Engines: 1.4 • 1.6 • 2.0 • 2.0D **MPG**: 34.9-46.3 **Gearbox:** Man/ Auto **Boot:** 440/490 **Seats**: 5 **Doors:** 3/4/5

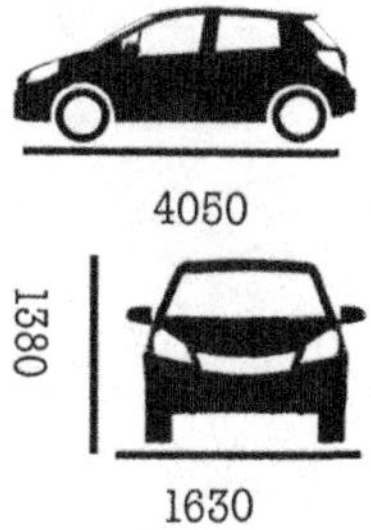

Peugeot 309 1987 - 1993

Brit designed hatch looking old fashioned in '90s, yet it remained a spacious and practical proposition. Small petrol struggles, 1.6 petrol is perfect and 1.8 diesel goes on forever delivering almost 50mpg. Neglected and underrated so has oil and coolant leaks, blown gaskets and smoky clattery diesels which make a racket as the body falls apart.

Engines: 1.1 • 1.4 • 1.6 • 1.9 • 1.8D • 1.9D **MPG**: 34.7-48.2
Gearbox: Man **Boot:** 295/615 **Seats**: 5 **Doors:** 3/5

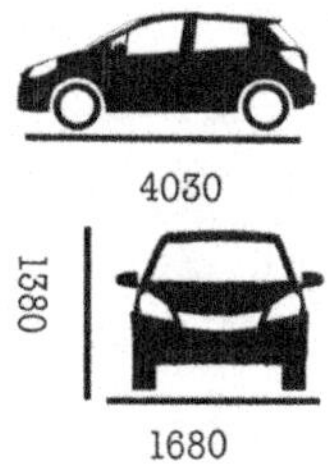

Peugeot 306 1993 - 2001

A Pug faced Citroen ZX. Looked even prettier from '97 and got safer with ABS and airbags a year later. There were still saloons around in the late '90s, but these are all about the diesels with hatchback bodies. Became even harder to find as the range was pared down to just estates and cabs. Worn suspension, driveshafts, plus a lot are very tired and rusty.

Engines: 1.4 • 1.6 • 1.8 • 1.9D • 2.0D **MPG**: 33.6-54.3
Gearbox: Man/Auto **Boot:** 338/637 **Seats**: 5 **Doors:** 3/5

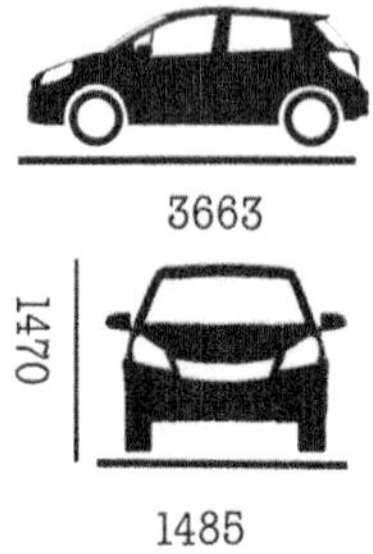

Renault 4 1961 - 1992

Legacy model with 8m editions, bought on price rather than as a retro thing. The 1.1 GTL was the ten year run out model which was able to keep up with modern traffic whilst making you deaf. Thinking Francophiles 2CV. Charming, even when disintegrating. Body rots away of course and the engine not entirely happy with unleaded fuel.

Engines: 1108 **MPG**: 45 **Gearbox:** Man **Boot:** 295/1175 **Weight:** 600 **Turning**: 10.1 **Seats**: 5 **Doors:** 3/5

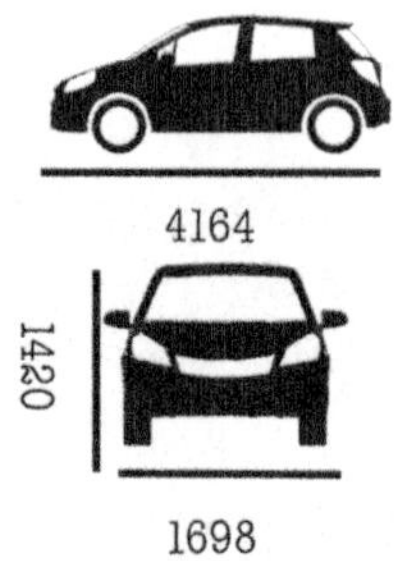

Renault Megane 1995 - 1999

Five body styles, sold well in the UK and proved themselves comfortable, safe in '90s terms, practical and cheap to run with model run out 1.4 and 1.6 engines had more fizz from '99. Complicated line up, its the RXE which delivers the most spec. Faults include the gearbox and electrics. Clutch repair difficult, coil packs and flywheel sensors can fail.

Engines: 1.4 • 1.6 • 2.0 • 1.9D **MPG**: 33.3-54.3 **Gearbox:** Man/ Auto **Boot:** 411/1237 **Seats**: 5 **Doors:** 3/4/5

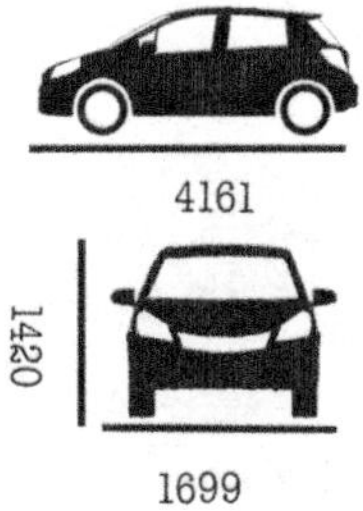

Renault Megane 1999 - 2002

Complicated line up with the less pretty second generation model. Loads of spin offs but we are sticking to the hatchback here. 1.4 or 1.6 petrol and the 1.9 diesels are the power packs behind what is one of the safest old bangers out there. Otherwise it is gearbox and electrics. Clutch repair difficult, coil packs and flywheel sensors.

Engines: 1.4 • 1.6 • 1.8 • 2.0 • 1.9D **MPG**: 36.7-54.3 **Gearbox:** Man/Auto **Boot:** 348/1210 **Seats**: 5 **Doors:** 3/4/5

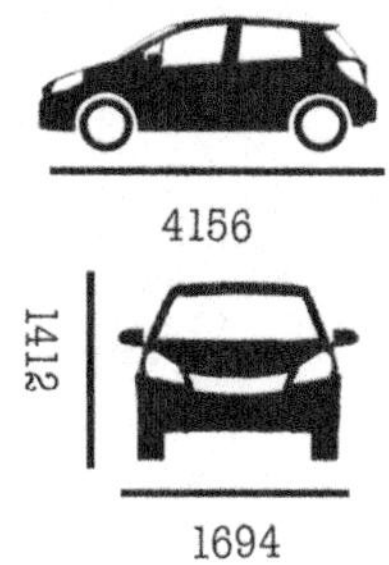

Renault 19 1988 - 1996

Decent family hatches and saloon (Chamade) that got much better with the Phase II in '92 with a restyle and some refreshed injection engines. Massive array of models, comfortable and obviously practical. Phase III in '93 brought a raft of upgraded models RN and RL Prima. Gear linkage can be weak, oil leaks, rough running and electricals.

Engines: 1.4 • 1.8 • 1.9D **MPG**: 35.3-46.1 **Kerb:** 948
Gearbox: Man/Auto **Boot:** 386-1310 **Seats**: 5 **Doors:** 3/4/5

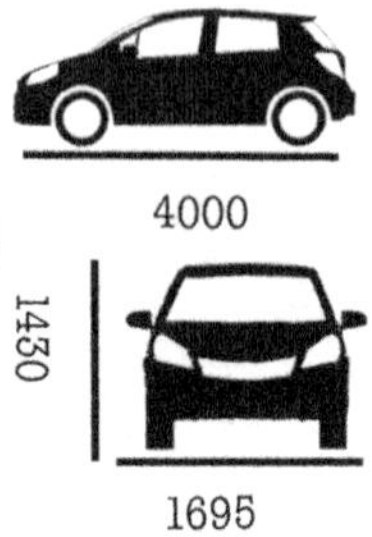

Rover Maestro 1990- 1994

Underwhelmingly modest effort which did most things OK. Old diesel delivered fairly spectacular 50mpg+ which made it a desirable low rent workhorse because there was loads of room inside. Simple lineup essentially Clubman and LX. '92 is when cats arrived and diesel turbo was an option. Water leaks, oil and coolant leaks, least of your worries.

Engines: 1.3 • 1.6 • 2.0D **MPG**: 39.0-55.6 **Turning**: 10.21
Gearbox: Man/Auto **Boot:** 300 **Seats**: 5 **Doors:** 5

Rover 200 1989 - 1996

Honda Concerto with a classy Rover twist which was a big hit with tough Peugeot diesels and smooth Japanese 1.6. SLis had power steering and is pick of the range. Felt modern at the time and got snazzy chrome grilles and side impact bars in '93. Rover engines can leak oil and sound cammy, Pug diesels get smoky and clattery.

Engines: 1.4 • 1.6 • 1.8 • 1.8D **MPG**: 32.3.-48.9 **Gearbox:** Man/Auto **Boot:** 300/610 **Seats**: 5 **Doors:** 3/5

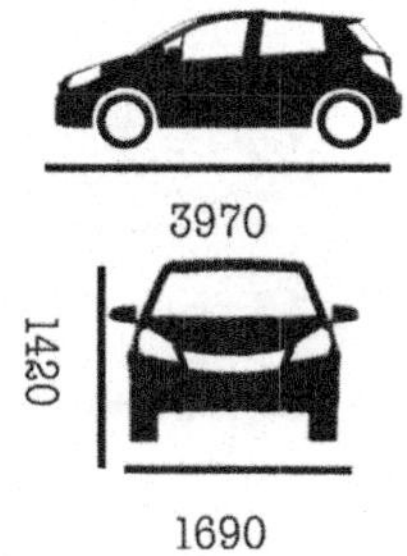

Rover 200 1995 - 1999

Full Rover and not half bad, being well built but a bit cramped in the back. Mature private buyers loved them and from '97 there was a leather and CD 216 GSi. There are quick ones and from '98 the entry level 211. Rattly camshafts, oil leaks, blown head gaskets, juddery clutches and leaking window seals.

Engines: 1.1 • 1.4 • 1.6 • 1.8 • 2.0D **MPG**: 38.4-50.1 **Gearbox:** Man/Auto **Boot:** 304/1086 **Seats**: 5 **Doors:** 3/5

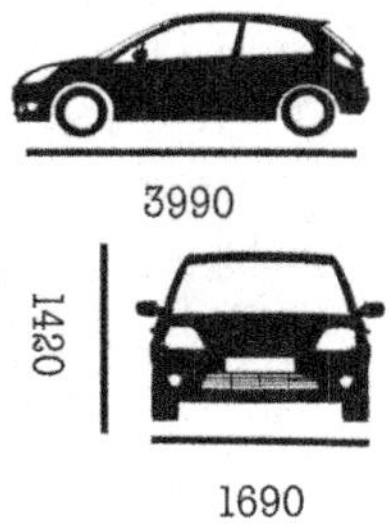

Rover 25 1999 - 2005

Rover 200, but with a smaller number. It hasn't got any bigger, simply rebadged. Notorious head gasket nonsense on the petrols might put you off and there is a gruff old diesel. Build quality can be questionable, so we are running out of reasons to buy. Three door is titchy really. Buy for old times sake only. Head gaskets, suspension wear, ABS +ECU issues.

Engines: 1.1 • 1.4 **CO2:** 160-164g/km **MPG**: 42.0-42.8 **Gearbox:** Man/Auto **Boot:** 304/1086 **Seats**: 5 **Doors:** 3/5

Sao Penza 1991 - 1993

Here's a BF-series Mazda 323 saloon and hatch, built in South Africa and imported to the UK by the Kia franchisee which meant in shared showroom space with the Pride, previously a Mazda 121. Just a cheapie import of the time which was only around for less than a couple of years. Minted mechanicals will outlast the dreary body.

Engines: 1.3 **MPG**: 39 **Speed:** 99 **0-60**:12.4**Weight**: 865 **Gearbox:** Man **Boot:** 315 **Seats**: 5 **Doors:** 4/5

SEAT Malaga 1985 - 1992

A Fiat Strada with a boot glued on is one description which also came with a 1.2 Italian petrol engine whilst much was made of the Porsche connection with the 1.5. The 1.7 diesel effectively replaced it in the UK. Not a lot to recommend it when it comes to being interesting. Sharp '80s styling, awful confused dashboard and a dull drive. Smoky worn engines.

Engines: 1.2 • 1.5 • 1.7D **MPG**: 34-42 **Weight**: 940 **Gearbox:** Man **Boot:** 515 **Seats**: 5 **Doors:** 5

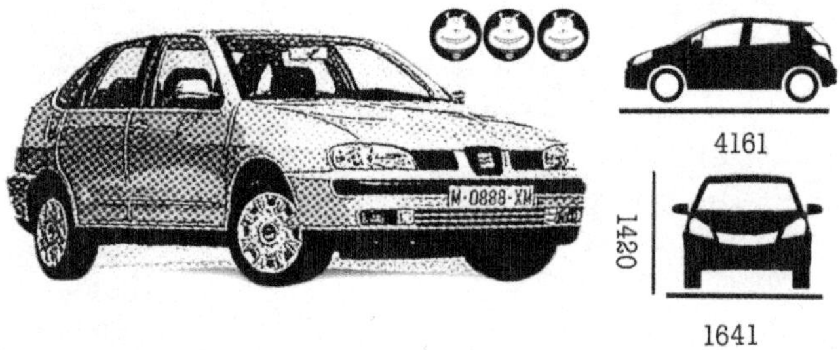

SEAT Cordoba 1994 - 2001

Polo related Spaniard with big old boot. Air con from '97, miracle if it is still working, the saloon and hatch are very practical load luggers. 1.9 diesel what an oil burner should be, otherwise 1.4 petrol is fine. Revamp in 1999 so models looked more alike. Oil leaks, suspension wear, mass air flow sensor failure. Rust making itself known.

Engines: 1.6 • 1.9D **CO2:** 135-189g/km **MPG**: 36.2-57.5 **Gearbox:** Man/Auto **Boot:** 455/762 **Seats**: 5 **Doors:** 5

Skoda Favorit 1988 - 1995

Czech new era that got better once the Volkswagen Group became involved. The Germans upped the quality, revised the suspension, interior and tweaked engines. Properly basic but from '94 alloys, removable rear seat and rear wash/wipe. L and LX marginal spec ones. Oily engines that get smoky, sticky gearboxes and poor electrical earths.

Engines: 1.3 • 1.6 • 1.9D **MPG**: 39.3-41.3 **Gearbox:** Man **Weight:** 894 **Boot:** 251/1038 **Seats**: 5 **Doors:** 5

Skoda Felicia 1994 - 2001

VW's Skoda spin that cured overheating issues and boosted the quality quotient to make the best budget buy of the '90s. Spacious and now with decent 1.6 and 1.9D VW engines. Restyled a bit in '98 and GLXi has glass sunroof, central locking, velour upholstery and a rev counter. 1.3 engine a bit creaky, worn steering, brakes and busted central locking.
Engines: 1.3•1.6•1.9D **MPG**: 38.2-49.8 **Gearbox:** Man **Weight**: 1058 **Boot:** 272/976 **Seats**: 5 **Doors:** 5

Suzuki Baleno 1995 - 2002

Utterly forgettable 1.6 saloon and three door hatch that despite revamps never became handsome. What it has going for it is Suzuki attention to detail and reliability. 1.8 engine also in the mix and not really much to recommend the Baloney except that it is a competent small family car. Might be a bit rusty now, just the usual wear and tear issues.
Engines: 1.6•1.8 **CO2:** 171-184g/km **MPG**: 36.7-39.8 **Gearbox:** Man/Auto **Boot:** 346 **Seats**: 5 **Doors:** 4

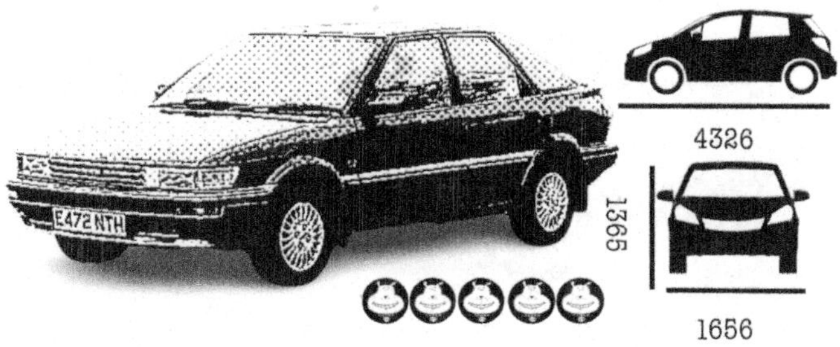

Toyota Corolla 1987 - 1992

6th generation brought lots of bodywork options, but the lift/ hatchbacks and saloons are the meat and two veg here. Actually a good car to own and drive and not the deadly dull thing history might have you believe. Power steering standard in 1990 which is fairly classy, unlike later incarnations. Driven into the ground, yet still survive.

Engines: 1.3 • 1.6 **MPG**: 32-40 **Gearbox:** Man/Auto **Boot:** 419 **Seats**: 5 **Doors:** 3/5

Toyota Corolla 1992 - 1997

Slightly less interesting than the last one, 3 door hatches and 5 door Lifts all very competent, but the 1.6 engine is the one which the best all rounder. An awful lot of specs to choose from the GXi has ABS, alloys and fog lamps. Revised in '95 and there's a ton of special editions, including the Kudos and Charisma. Last forever.

Engines: 1.3 • 1.6 • 1.8 • 2.0D **MPG**: 34.6-41.9 **Gearbox:** Man/ Auto **Boot:** 400/1120 **Seats**: 5 **Doors:** 3/5

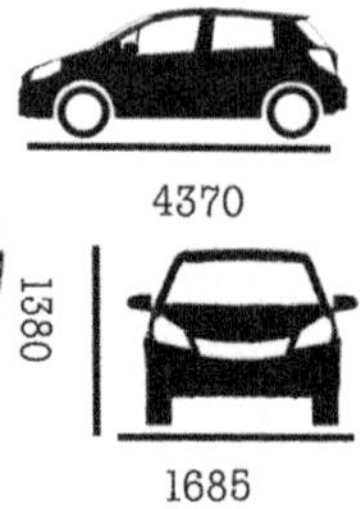

Toyota Corolla 1997 - 2002

Legend is an overused word, but the Corolla is so much more. A worldwide used car phenomenon. Liftbacks, saloons and all that with unbreakable petrols plus a diesel. Avensislike styling from 2000, saloon deleted and common rail diesel added. Made in Blighty. Rather rusty now, lambda sensors fail eventually, overall just worn out bits.

Engines: 1.4 • 1.6 • 2.0D **CO2:** 156-168g/km **MPG**: 40.4-48.7 **Gearbox:** Man/Auto **Boot:** 289/960 **Seats**: 5 **Doors:** 3/5

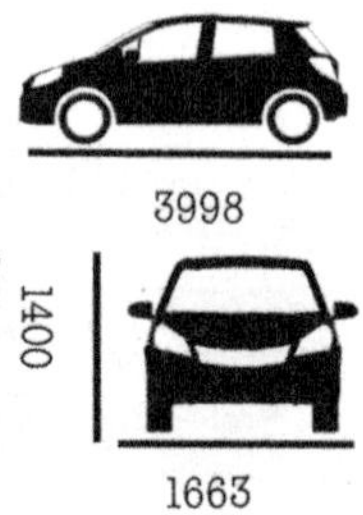

Vauxhall Astra 1984 - 1991

Arguably better than a contemporary Escort, blobby aero styling and it gets car of the year. Belmont saloon from '86. Great range of engines and the slogging diesel delivered amazing economy for the low level rep. A popular company and family motor which was cheap to fix and easy to look after. Worn, brakes, suspension and smoking diesels.

Engines: 1.2 • 1.3 • 1.4 • 1.6 • 1.8 • 2.0 • 1.6D • 1.7D **MPG**: 34-48 **Gearbox:** Man/Auto **Boot:** 360/720 **Seats**: 5 **Doors:** 3/4/5

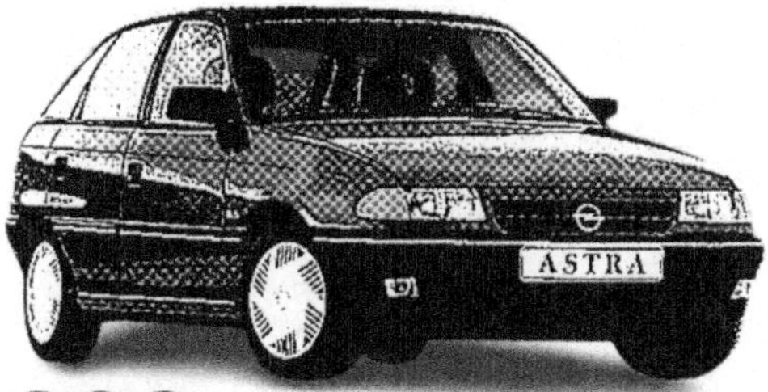

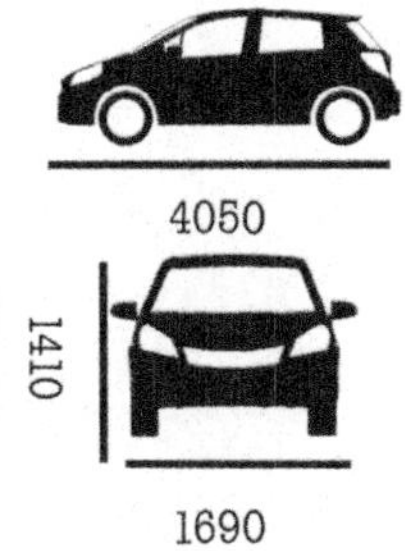

Vauxhall Astra 1991 - 1998

Very much more of the same and quite right too, dated but reliable. 16 valve engine a lot more lively, but the dreary 8v 1.4 will get you there and the grim 1.7D without using much fuel at all. August '94 the V grilles with power steering, uprated suspension and other mods. CDX from '95 has alloys and CD. Worn engines, clutch and suspension.

Engines: 1.4 • 1.6 • 1.8 • 2.0 • 1.7D **MPG**: 34.1-49.1 **Gearbox:** Man/Auto **Boot:** 360/1200 **Seats**: 5 **Doors:** 3/4/5

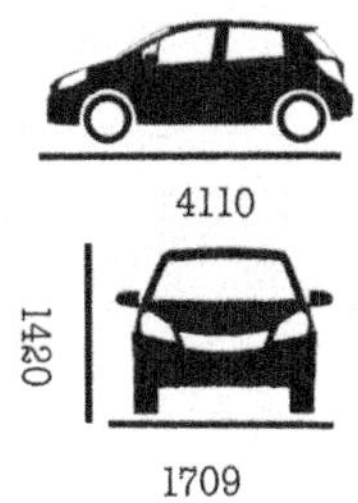

Vauxhall Astra 1998 - 2005

Mark 4, quite possibly the best Astra in all of human history. Massive range, seven engines and extremely roomy inside. Easy to drive and live with, but open to company car abuse. Quite tough in the real used car world, underrated which keeps prices marginal. Corroded ECUs, cockpit leaks, failing Air conditioning and water pumps.

Engines: 1.4 • 1.6 • 1.8 • 2.0 • 1.7D • 2.0D **CO2:** 154-172g/km **MPG**: 38.2-49.6 **Gearbox:** Man/Auto **Boot:** 370/1210

Volkswagen Beetle 1999 - 2010

Here's a Golf in a frumpy dress. Retropointless as there is less room and mpg than the original. Bought as a last resort, or as a personal transportation pod, either slow 1.6, or amusing V5. Stick to the 1.9 diesel for all round usefulness and minimal pointlessness. Lots of minor electrical failures, windows, locks etc. Engine ECU. Now bubbling rust.

Engines: 1.4 • 1.6 • 1.8 • 2.0 • 2.3 • 1.9D **MPG**: 36.7-53.3
Gearbox: Man/Auto **Boot:** 209/769 **Seats**: 4 **Doors:** 3

Volkswagen Golf 1984 - 1992

Many might argue the bestest Golf of all time, heavy to drive without power steering, especially GTI but solid as a rock. Avoid the tiniest petrols, automatics good, diesel delivers a solid mpg return. Mean specifications, needs to be at least GL. Can have oil leaks, blown head gaskets, smoky diesels, snapped clutch cable. Mostly neglect, overall tough as nuts.

Engines: 1.0 • 1.3 • 1.6 • 1.8 • 1.6D **MPG**: 35.1-54.2 **Gearbox:** Man/Auto **Boot:** 345 **Seats**: 5

Volkswagen Golf 1992 - 1998

Blobby styling and not that well put together, the only lemon shaped Golf. Despite the moans another massive hit. Constantly updated and the bigger petrols engines always better, GTI disappoints, everything is average but has ultimately proved quite durable. Unloved so smoky diesel, clattery petrols and general wornoutness.

Engines: 1.4 • 1.6 • 1.8 • 2.0 • 1.9D **MPG**: 35.4-56.8 **Gearbox:** Man/Auto **Boot:** 330/1162 **Seats**:5

Volkswagen Golf 1997 - 2004

VW had something of a wobble with the last model, this one was spot on when it came to fit and finish and although quite old, have held up. Buy a five door boring 1.9 diesel or an interesting V5 three door. Hard to go badly wrong, but the GTIs weren't all that. Rust on sills and arches. Engine ECU cause rough running, brake calipers and water pumps.

Engines: 1.4 • 1.6 • 1.8 • 2.0 • 2.3 • 1.9D **MPG**: 32.1-57.6 **Gearbox:** Man/Auto **Boot:** 380/1270 **Seats**: 5 **Doors:** 3/5

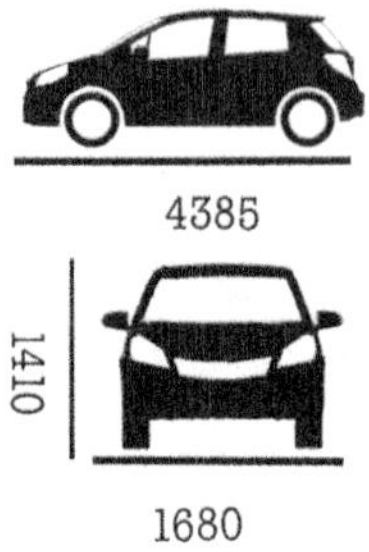

Volkswagen Jetta 1984 - 1992

Officially a Jetta II and in reality a Golf with a boot. This is a Golf clone with more of a market in Europe and elsewhere, always the value used buy in Blighty. Basically as Golf then so the same applies when it comes to going for the bigger engines. Can have oil leaks, blown head gaskets, smoky diesels, snapped clutch cable. Tough as booted nuts.

Engines: 1.3•1.6•1.8•1.8D **MPG**: 35.1-54.2 **Weight**:910
Gearbox: Man/Auto **Boot:** 575 **Seats**: 5 **Doors:** 4

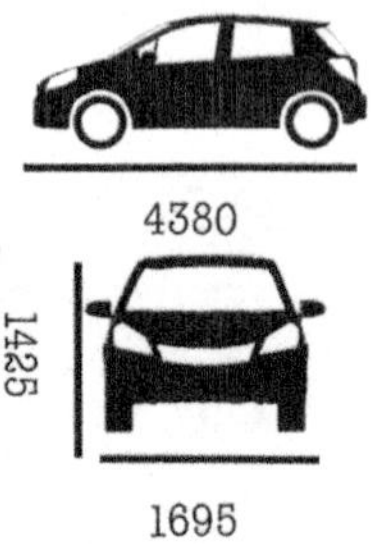

Volkswagen Vento 1992 - 1998

Changing the name doesn't change what it is, a booted Golf of course. Main attraction always was that its cheaper than the equivalent Golf, and actually these were better built and felt more solid. Side airbags and airbags from '94. Even more unloved than the Golf 3 so smoky diesel, clattery petrols and general wornoutness.

Engines: 1.6•1.8•2.0•1.9D **MPG**: 34.1-55.3 **Weight**; 1045
Gearbox: Man/Auto **Boot:** 527 **Seats**: 5 **Doors:** 4

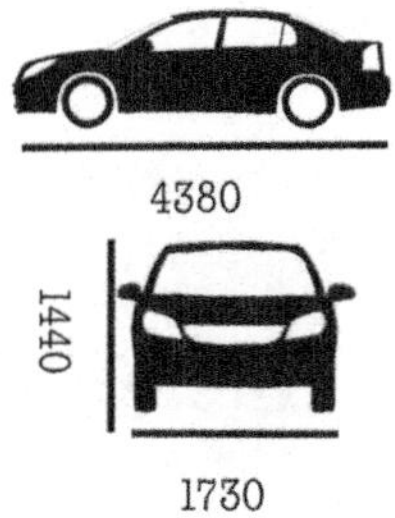

Volkswagen Bora 1999 - 2005

The Golf with a boot never went away. So it is less practical and more boring looking, that's the trade off. Mark 4 Golf is a rather good basis for something so conventional. V5 and V6 not very conventional the 1.9TDi restores the buying balance. Rust on sills and arches. Engine ECU cause rough running, brake calipers and water pumps.

Engines: 1.6 • 1.8 • 1.9D **CO2:** 143-192g/km **MPG**: 39.8-53.3
Gearbox: Man/Auto **Boot:** 455/785 **Seats**: 5 **Doors:** 4

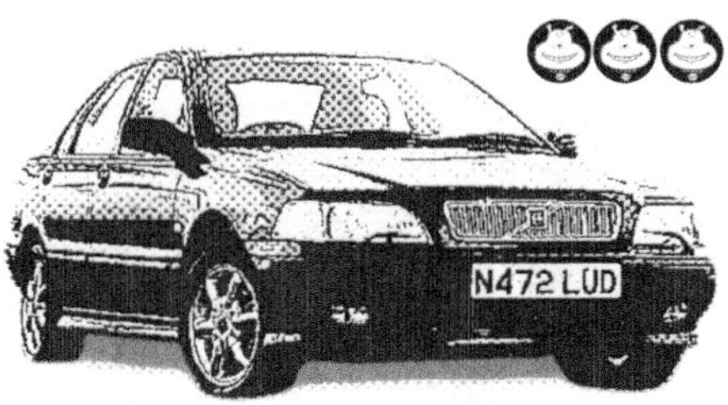

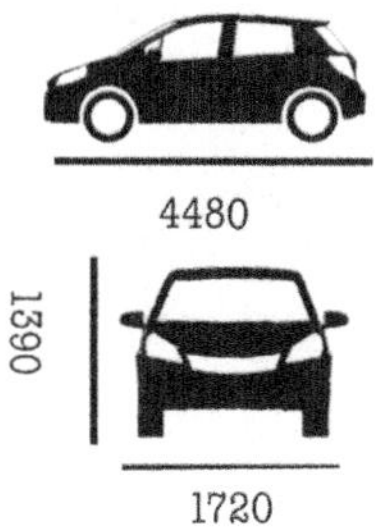

Volvo S40 1996 - 2004

Most people bought the estate. Not that special and can be neglected. Mitsubishi Lancer related, so tough as they come if not the high quality saloon you might expect, just a dull worker. Suits the 1.9D engine, bodywork starts to rot. Corroded loom causes poor electrics, air con/power steering can fail, brake calipers break.

Engines: 1.6 • 1.9D **CO2:** 142-185g/km **MPG**: 35.8-52.3
Gearbox: Man/Auto **Boot:** 415 **Seats**: 5 **Doors:** 4

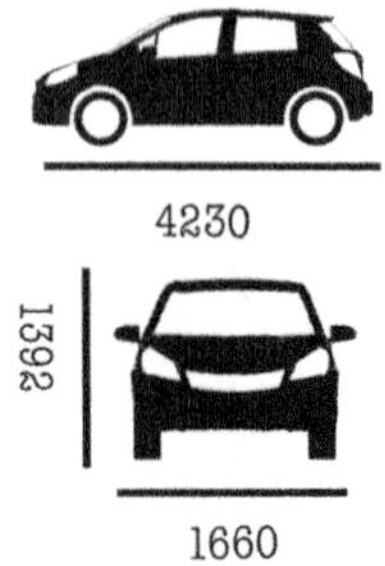

Volvo 360 1982 - 1991

In its twilight year the range was wound right down to the 5 door hatch. Beloved of their mostly elderly owners who appreciate the van size carrying capacity and don't mind Renault power. There's a huge number of what to look fors from the cracked manifold to the CVT transmission playing up, water pump and much more. Solid though.

Engines: 1.4•1.7•2.0 **MPG**: 32.2-43.6 **Weight** 1091 **Gearbox:** Man/Auto **Boot:** 1800 **Doors**: 3/5

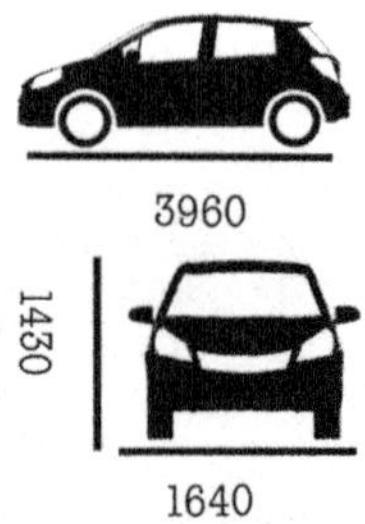

Zastava Yugo Sana 1989- 1993

Another Eastern blocker trying to come to terms with the current decade. This is a reheated Fiat Tipo underneath with an OK ItalDesign body on top. Fairly terrible as a package despite the Tipo ingredients, so not good drive or great to own, built down to a price, but had a rear wash/wipe and a rev counter, as if that mattered. Grim as they come and go.

Engines: 1.4 **BHP**: 70 **MPG**:43 **Gearbox:** Man/Auto **Boot:** 410/1280 **Weight**:873 **Seats**: 5 **Doors:** 5

Ex Fleet Family Favourites

Here is the Meat and Two Veg of the Bangerpedia, those ex-company cars that make up the bulk of the used car market. As ever, what a time to be alive and use some of these beauties as your everyday workhorse.

Of course the Execubangers in a few pages time were great, but these standard issue, regional assistant sales transportation devices were better than anyone really gave them credit for. Although the Sierra limped into the '90s, the Mondeo was a class apart and not just great to drive, but slightly stylish too. The revamped Cavalier took the motorway slog very much in its stride, then the Vectra turned up as a massively accomplished muncher of miles. The Rover Montego was dated and the 400 was OK, but the real surprise was how good the Passat became. Always a decent Banger it now had a certain style towards the end of the '90s.

French cars did the comfy thing, but only the 405 and 406 were truly inspiring. Hats off to Citroen for continuing to make technically intriguing vehicles but that wasn't what you wanted on the hard shoulder of the M6. Actually what you wanted was a British built Toyota Carina E that would roll on forever. The Avensis was a bit more dreary, but filled the reliability gap. Speaking of reliability and dullness the Nissan Primera was always there or thereabouts when you needed to put the miles in. Like many other cars of the era a British Touring Cars star and there was always a hot halo version, the 2.0e GT, to make the range seem a bit more relevant.

There was always a lot of middle order dross, priced to be attractive that turned out to be prime Banger fodder within minutes of being sold new. The Proton Persona and Mpi were certainly nothing special. Meanwhile the Daewoo Espero and Nubira were trying to do something a bit different at your local Halfords outlet.

These were at least were interesting times.

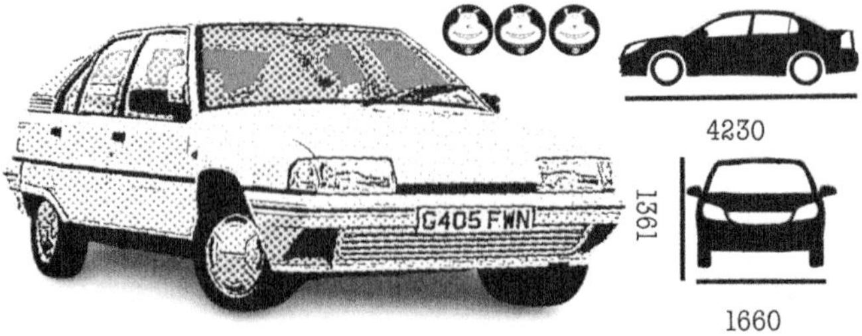

Citroen BX 1982 - 1994

The Phase II models from 1990 is effectively the run out ones with an exciting 4×4 introduced. Super comfy, practical and used to be cheap because of the complexities. The diesels are old school but great. Plenty to go wrong, suspension and steering, camshaft wear, snapped clutch cables, general electric upsets. Remains more of an '80s icon.

Engines: 1.6 • 1.9 • 1.7D • 1.9D **MPG**: 33.8-48.2 **Gearbox:** Man/Auto **Boot:** 444/1455 **Seats**: 5 **Doors:** 5

Citroen Xantia 1993 - 2000

'90s soft wedge and one of the very last mass market interesting Citroens. Hydractive suspension is great for comfort and can be terrible when its a FrancoBanger. Otherwise lovely to drive, great engines from V6 to the turbo diesels. Phase 2s in '98 which are stronger and safer. Apart from the hydraulic related issues, ECUs, diesel injectors.

Engines: 1.6 • 1.8 • 2.0 • 3.0 • 1.9D • 2.1D **MPG**:24.4-44.7 **Gearbox:** Man/Auto **Boot:** 480/878 **Seats**: 5 **Doors:** 5

Daewoo Espero 1995 - 1997

If you liked the old Cavalier saloons from the '80s, you'll love this. Safer with airbag and ABS, better specced too, CDXi has alloys, not all had air. GLXi came with on board mobile phone. Unremarkable overall, sold on value rather than much else but reliable enough. Leaking window and sunroof seals, noisy camshafts, air con can be a bit flaky.

Engines: 1.5 • 1.8 • 2.0 **MPG**: 34.7-35.9 **Gearbox:** Man/Auto
Weight:1088 **Boot:**560 **Seats**: 5 **Doors:** 4

Daewoo Nubira 1997 - 2002

Not a warmed up Vauxhall anymore, but not anything to get excited about. Europeanised in GB for the brave new big box dealership world they were trying to invent. OK to drive, overall quality just about OK. Got a revamp in '99 but mainly styling. Good basic specification, but you need to get a CDX for all round disc brakes. Problems: mostly wear and tear.

Engines: 1.6 • 2.0 **MPG**: 29.7-32.1 **Gearbox:** Man/Auto
Weight: 1165 **Boot:** 370/890 **Seats**: 5 **Doors:** 4/5

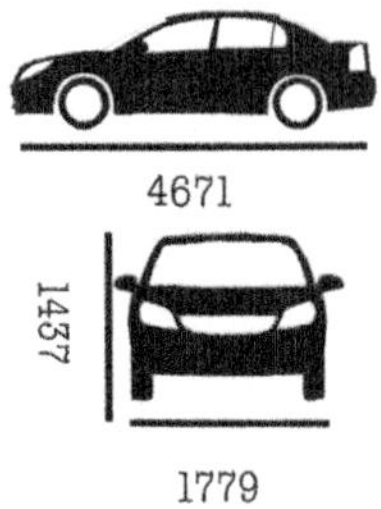

Daewoo Leganza 1997 - 2002

These are almost extinct now, just a 2.0 petrol saloon with bucket loads of standard kit. It looks like a very low rent Lexus GS, but isn't that refined to drive. Lots of room and on the whole reliable. Never held its value, so pay a £1 and drive it into the ground. Nothing to look for, just wear and tear items like brakes and suspension.

Engines: 2.0 **CO2:** 215g/km **MPG**: 30.7 **Gearbox:** Man/Auto **Boot:** 400 **Seats**: 5 **Doors:** 4

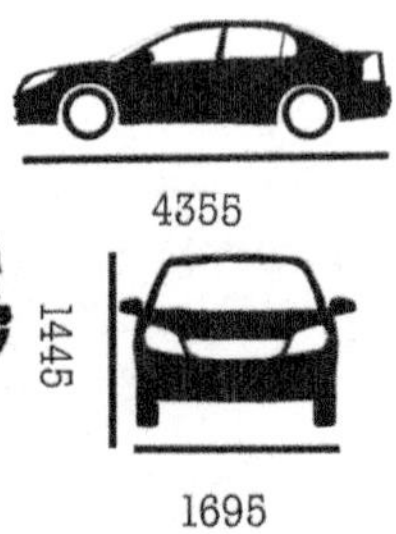

Fiat Tempra 1990 - 1996

Tipo with a great big boot. Perfect family car, but few sold as the dullness put most off. Cheap used and therefore abused. Actually good to drive. Got safer from '93 with a stronger body, side impact bars and airbag. S, SX and SLX, avoid the basic ones, but the SX has digital instruments that may have gone AWOL. Worn valve guides, iffy gearboxes and CVT.

Engines: 1.4•1.6•1.8•2.0•1.9D **MPG**: 33.7-47 **Gearbox:** Man/Auto **Weight**:1075 **Boot:** 498 **Seats**: 5 **Doors:** 4

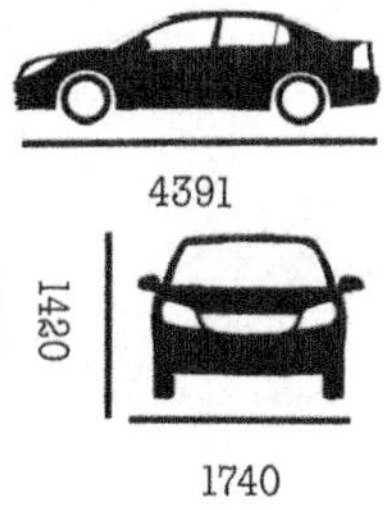

Fiat Marea 1997-2002

Stylish thing which is nice to drive, but the fact it is a Fiat meant no one took it seriously new or used. Petrol and diesel engines complemented by not very many standard bits of kit. If any are around, should cost buttons. Chocolate ones. Suspension drop links, sunroof leaks, instrument panel failures, needs red master key otherwise worthless.

Engines: 2.0 • 1.9D **CO2:** 149-234g/km **MPG**: 28.8-50.4 **Gearbox:** Man/Auto **Boot:** 438 **Seats**: 5 **Doors:** 4

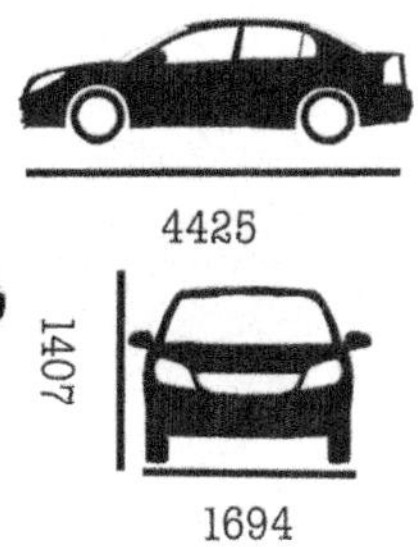

Ford Sierra 1982- 1993

Jellymould a bit creaky by now, but still wildly popular as a hand me down from company fleets to families. It is like working on an old Cortina which makes it one of the better recent period Bangers. Loadsa specs and specials, but ok from the LX upwards. Wear and tear which is all fixable apart from terminal bodywork.

Engines: 1.6 • 1.8 • 2.0 • 2.9 • 1.8D **MPG**: 26.1-38.7 **Gearbox:** Man/Auto **Boot:** 420 **Seats**: 5 **Doors:** 5

Ford Sierra Sapphire 1989- 1993

Best known for the mad Cossie incarnation, otherwise its a dated rep special and prime Banger with limited appeal. Loadsa specs and specials, but OK from the LX upwards, but you deserve a Ghia. 2.0 is the best all rounder, diesels are plodders. Wear and tear, worn cams and valve guides which is all fixable apart from terminal bodywork.

Engines: 1.6 • 1.8 • 2.0 • 1.8D **MPG**: 28-38.7 **Gearbox:** Man/Auto **Boot:** 413 **Seats**: 5 **Doors:** 4

Ford Mondeo 1993 - 1996

Ford's great leap forward and creation of Mondeo Man. Not a looker, but great to drive, practical, comfortable. Like the Sapphire needs to be GLX and above, whilst the Ghia X has knobs on like cruise, leather and a CD. Four wheel drive again and there's a sporty Si with revised suspension. Can have a sticky clutch and worn suspension. Mostly just wear.

Engines: 1.6 • 1.8 • 2.0 • 2.5 • 1.8D **MPG**: 30.6-46.3 **Gearbox:** Man/Auto **Boot:** 470/965 **Seats**: 5 **Doors:** 4/5

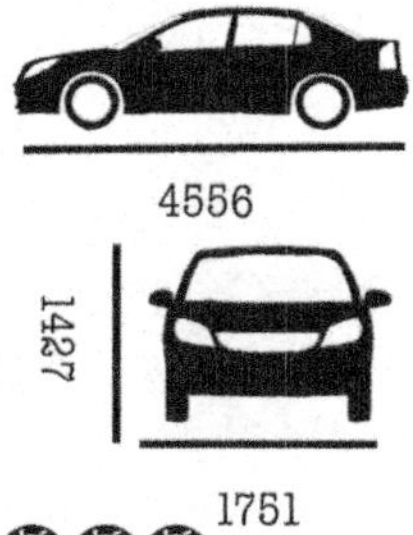

Ford Mondeo 1996 - 2000

Enough of an update to warrant Mondeo Mark II status with a new face plus uprated suspension, engines, gearbox and the interior. Good standard spec above the Aspen and Verona, so air CD system and Quicklear windscreen. More to go wrong, but mostly it doesn't if looked after, but multi owners and mega mileages eventually wore them out.

Engines: 1.6 • 1.8 • 2.0 • 2.5 • 1.8D **MPG**: 30.6-46.3 **Gearbox:** Man/Auto **Boot:** 470/965 **Seats**: 5 **Doors:** 4/5

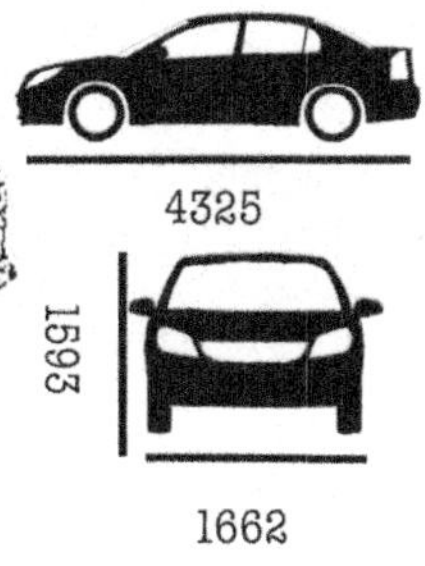

Hindustan Ambassador 1992 - 1998

Actually its a Fullbore Mark 10. 1.8l, 74bhp Isuzu engine remained the same, but the paintwork was re-sprayed, heaters and catalytic converters were installed, along with new weather seals, tyres and a front anti-roll-bar. Pointless instaretroclassic exercise that few fell for. Curiosity value only. Drum brakes all round...

Engines: 1.8 • **MPG**: 26.4 **Gearbox:** Man Weight: 1150 **MPH;** 90 **Seats**: 5 **Doors:** 4

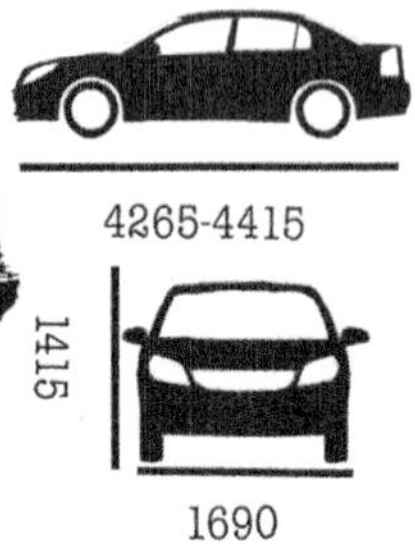

Honda Concerto 1991 - 1995

Here is Honda's Rover 200, or is it the other way around? You'll find the 200 in the previous section for some reason, the Concerto is another old codger favourite, reliable and unassuming. Revised in '93 with modded suspension and new interior trim. Honda engines something of a long lasting treat compared to a flawed K-Series.

Engines: 1.5 • 1.6 **MPG**: 32.3-35.1 **Gearbox:** Man/Auto **Boot:** 370/1210 **Weight:** 1070 **Seats**: 5 **Doors:** 4/5

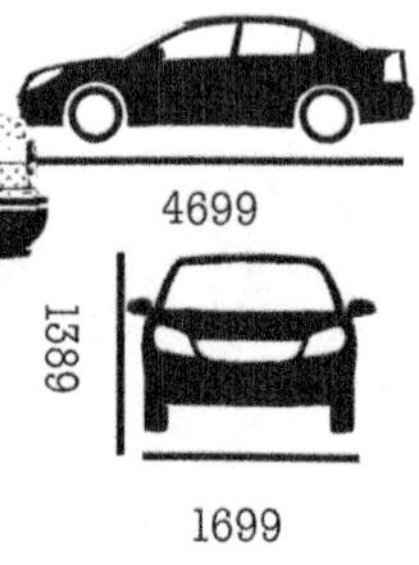

Honda Accord 1989 - 1993

Long forgotten classy range of unbreakable middle order vehicles that were comfortably dull. Not a great deal to say about these apart from the obvious, which is the cost of repairing one these days. The uninjected 2.0 is probably worth avoiding, otherwise there is an outside chance of a blown head gasket, worn gearbox and leaky power steering.

Engines: 2.0 • 2.2 **MPG**: 29.1-32.7 **Gearbox:** Man/Auto **Boot:** 425 **Turning**: 10.8 **Seats**: 5 **Doors:** 4

Honda Accord 1993 - 1999

Here's a true Brit for you, built in Blighty but has Honda badges. Better than the 600 in value terms because of the spec, long term it seems to be the more mechanically intact survivor. Range revised in 1998 and the Executive provides comfort and stylish fittings. Tough, not as chintzy as a 600, but that's not why you buy one of these. Solidly sensible.

Engines: 1.8 • 2.0 • 2.2 • 2.0D **MPG**: 31.4-54.9 **Gearbox:** Man/Auto **Boot:** 485-605 **Seats**: 5 **Doors:** 4/5

Honda Accord 1999 - 2003

Fabulous family transport which is big enough, nice to drive and built to such a high standard. This is more Banger royalty. No diesel, just superb petrol engines that will go on and on. Rust is the only enemy and serious owner neglect. It really is the wear and tear items like brakes and suspension. Some parts though may cost a bit or be difficult to find.

Engines: 1.8 • 2.0 • 2.3 **CO2:** 203-209g/km **MPG**: 32.1-32.8 **Gearbox:** Man/Auto **Boot:** 430 **Seats**: 5 **Doors:** 4

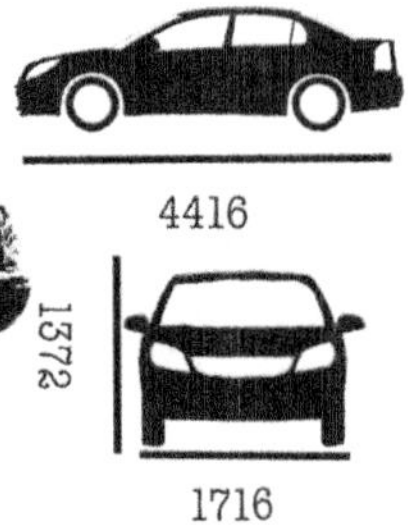

Hyundai Stellar 1984 - 1992

Whatever happened to the Ford Cortina? Well essentially this was better than the original. Built the Dagenham Dustbin under licence, then came up with this which had Mitsubishi mechanicals. Perfect Banger that would last several lifetimes. 1600 GSL had it all, good old fashioned motoring values, that could always be fixed.

Engines: 1.6 **MPG:** 33 **Gearbox:** Man **Boot:** 433 **Weight:** 1025 **Seats**: 5 **Doors:** 4

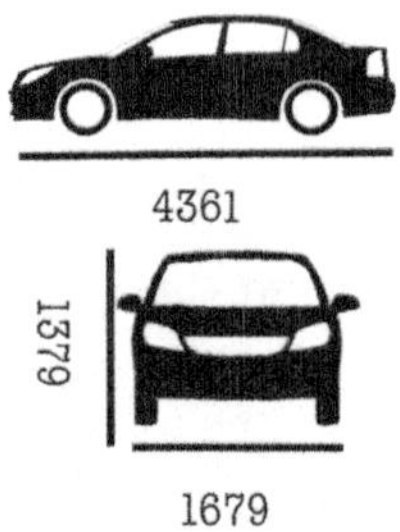

Hyundai Lantra 1991 - 1995

Building block Banger that helped established the Koreans as major payers in the modern motoring world. An unremarkable saloon that had a decent amount of kit especially in the GLSi and CDi and the toughest of mechanicals, because they are Mitsubishi sourced. Smoke and rattles from old engines and early suspension fractures.

Engines: 1.5•1.6•1.8 **MPG:** 29-41 **Gearbox:** Man/Auto **Boot:** 331 **Weight:** 1017 **Seats**: 5 **Doors:** 4

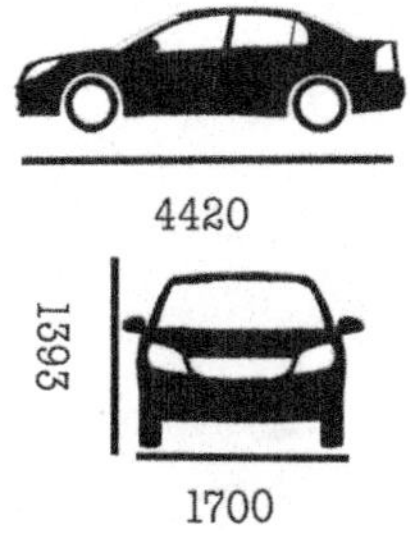

Hyundai Lantra 1995 - 2000

Hyundai hitting its stride now and replacing the older Mitzi engines. A bit dull still, not that huge inside, but largely generous spec with CD and GSi. Seem to be immune to basic breakdowns because of the solid engineering, so the overall underwhelming experience is understandable. Smokey and rattly engines when neglected and old.

Engines: 1.6 • 1.8 • 2.0 **MPG:** 33.2-34.9 **Gearbox:** Man/Auto **Boot:** 394 **Weight**: 1068 **Seats**: 5 **Doors:** 4/5

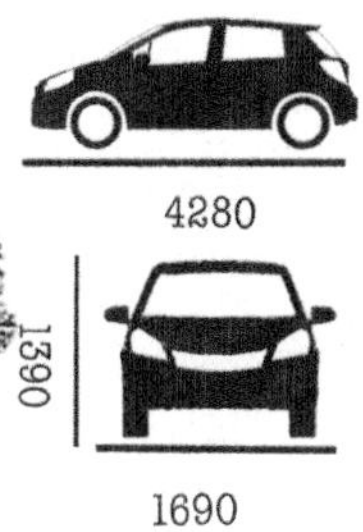

Kia Mentor 1994 - 2003

Just like the Escort, this just makes it into the Bangerpedia and is a similar sort of depressing '90s saloon. Even so it was based on the more interesting Mazda 323, which guarantees it won't break. Became a Mentor II then a new Mentor in 2001. That's enough Mentor. Nothing to see there, just the usual worn suspension and brakes.

Engines: 1.6 **CO2:** 190g/km **MPG**: 35.3 **Gearbox:** Man/Auto **Boot:** 328 **Seats**: 5 **Doors:** 4

Kia Clarus 1999 - 2001

It is just a great big cheapo pretend execubarge with plastic wood trim. MPG is not a strong point, spec certainly is even with the basic SX, whilst t he top of the line Executive is all leather and CD. It's a Kia so the reliability is not exactly an issue. Largely a big softy and only neglect is an issue so worn out bits like the exhaust and suspension..

Engines: 1.8 • 2.0 **MPG**: 27.4-30.4 **Gearbox:** Man/Auto **Boot:** 354/764 **Weight**: 1209 **Seats**: 5 **Doors:** 4

Mazda 626 1987 - 1991

Here's the thing, a four wheel drive, four wheel drive proto-super saloon in GT trim. There were Billy Basic ones, but here was a possible future involving a fairly nippy, but not cheap, saloon and hatch that could do 130mph. It's an odd concoction so when it went wrong then it was rarely repaired properly or at all. Worth finding.

Engines: 1.8 • 2.0 **MPG**: 25 **Gearbox:** Man/Auto **Boot:** 490 **Weight**: 1125 **Seats**: 5 **Doors:** 4/5

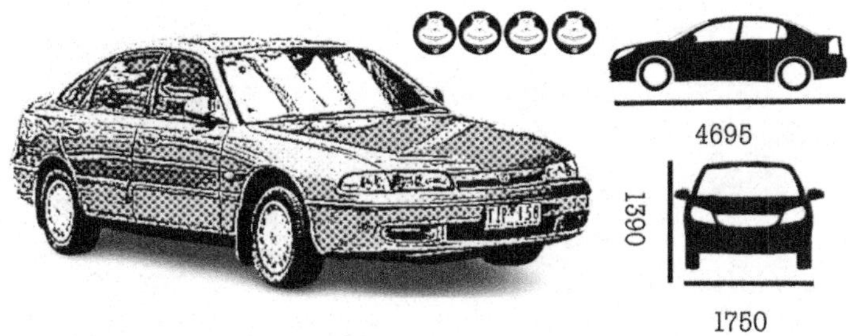

Mazda 626 1992 - 1997

Dumbed down compared to the old one, but there is a V6 for entertainment value. It is though a well equipped and solidly put together middle order motor. GLX delivers most of the spec you need including electric door mirrors whilst an Executive adds velour, plus a touch of walnut and cruise. Not much goes wrong, if you can find one.

Engines: 1.8 • 2.0 • 2.5 • 2.0D **MPG**: 29.3-41.1. **Gearbox:** Man/Auto **Boot:** 452 **Weight**: 1150 **Seats**: 5 **Doors:** 4/5

Mazda 626 1997 - 1999

Here is the interim model before the 626 really hit the big time with company fleet buyers. Specification upgraded periodically. Very much an anti-climax of a car, does everything averagely well, but what it excels at is not breaking down. Top notch SE delivers alloys, cruise and some pointless wood trim.

Engines: 1.8 • 2.0 • 2.0D **MPG**: 34.9-54.3 **Gearbox:** Man/Auto **Boot:** 502 **Weight**: 1150 **Seats**: 5 **Doors:** 4/5

Mazda 626 1999 - 2002

Scaling up the 323 worked, here was a dull, but very worthy saloon and hatchback. The restyle in 1999 was meant to help but it was still dreary. It's an old school minicab because it is roomy enough and needs just basic servicing to keep on working, probably forever. These are remarkably tough, just neglect and worn out parts such as suspension and brakes are the killers.

Engines: 1.8 • 2.0 • 2.0D **CO2:** 152-195g/km **MPG**: 35.8-54.3 **Gearbox:** Man/Auto **Boot:** 502 **Seats**: 5 **Doors:** 4/5

Mazda Xedos 6 1992 - 1999

Attempt to Posh Up the brand with a V6. Available in SE trim which added leather. Pauper four cylinder 1.6 arrived in '93, then in '94 uprated suspension, alloys and fogs and ordinary 1.6 now called Sport as a manual only. Largely forgotten, a bit tight inside, but the trade off is ultra Mazda reliability. Wear and tear and decades take their toll.

Engines: 1.6 • 2.0 **MPG**: 33.7-36.8 **Gearbox:** Man/Auto **Boot:** 407 **Weight:** 1220 **Seats**: 5 **Doors:** 4

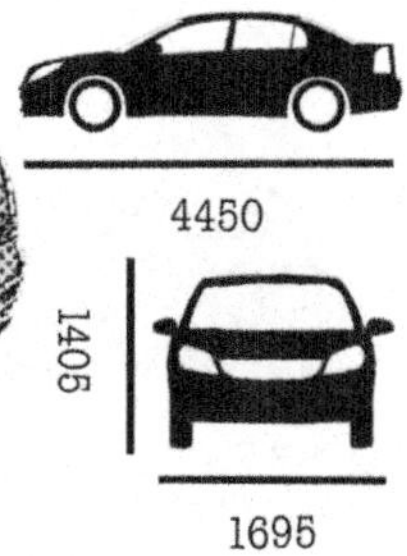

Mitsubishi Carisma 1996 - 1999

One if the most anonymous rep specials related to the Volvo S40 but considerably duller. Dutch build as good as the Far Eastern. Find a Diamond pack model and get ABS, electric sunroof and a passenger airbag. Otherwise GLS is the top model with alloys and colour keying, plus there are special editions. Not much happens to these. Has it got two keys?

Engines: 1.6 • 1.8 • 1.9D **MPG**: 38.4-46.9 **Gearbox:** Man/Auto **Weight:** 1105 **Boot:** 430 **Seats**: 5 **Doors:** 4/5

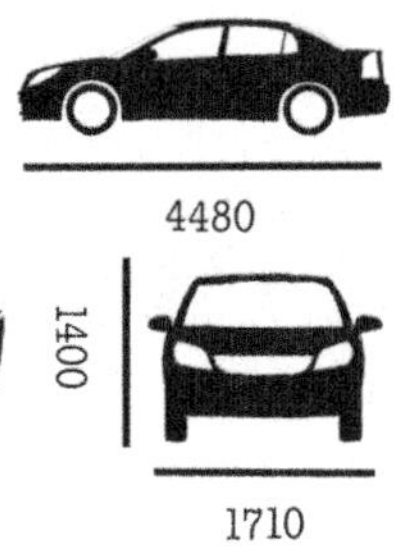

Mitsubishi Carisma 1999 - 2004

The perennial gag is that it doesn't have any. Another of those anonymous '90s Japanese hatches and saloons, which just got on with working. Lots of standard equipment with a Classic, Equippe, Elegance or Sport. Low running costs make up for lack of personality. Not much to look for, just wear and tear, incredibly fault free.

Engines: 1.6 • 1.8 • 1.9D **CO2:** 168-176g/km **MPG**: 38.2-46.3 **Gearbox:** Man/Auto **Boot:** 430 **Seats**: 5 **Doors:** 4/5

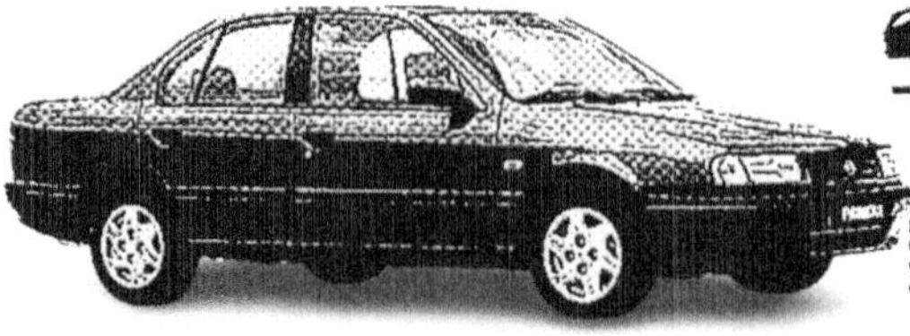

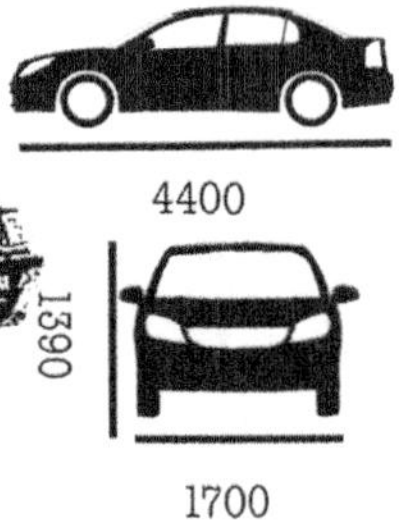

Nissan Primera 1990 - 1996

It could not be less spectacular, but that is why it's spot on. A fleet car legend which got on with the job, by not breaking down then becoming THE minicab of choice. Series 2 in '93 with side impact beams and more equipment. SE has leather, alloys, wood and CD. Diesel not that frugal, whereas the 1.6 was. Survivor of extreme wear and tear and neglect.

Engines: 1.6 • 2.0 • 2.0D**MPG**: 33.4-39.1 **Gearbox:** Man/Auto **Weight**: 1160 **Boot:** 480/1000 **Seats**: 5 **Doors:** 4/5

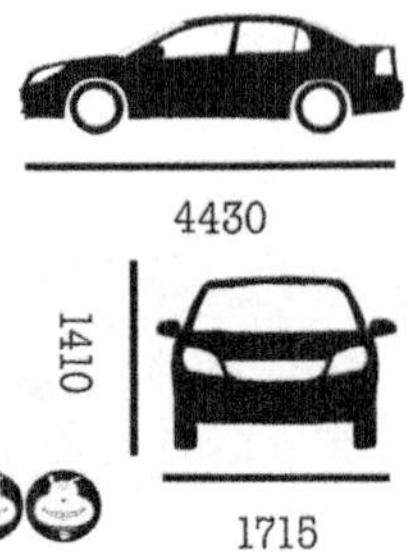

Nissan Primera 1996 - 1999

Second Gen, the P11 in model code terms, still looks dull, but otherwise was sharpened up in the handling department. Dynamically great, mechanically sound with better engines, petrols are great, diesel is just diesely. Decent equipment from the Equation onwards. SLX, SE are the top specs, and trip computer on the GT and SRi. Will be worn out.

Engines: 1.6 • 1.8 • 2.0 • 2.0D**MPG**: 34.8-42.2 **Gearbox:** Man/Auto **Weight**: 1215 **Boot:** 490/1000 **Seats**: 5 **Doors:** 4/5

Nissan Primera 1999 - 2002

Here was the UK's favourite minicab in the '90s based on the fact that it would just keep on working, probably forever. Lots of standard kit of course. The engines are great and it is only going to be consumables like tyres and exhausts that may need replacing. Older examples are starting to rot, suspension wear, air con busted, old alloys corrode.

Engines: 1.6 • 1.8 • 2.0 • 2.0D **MPG**: 34.8-43.5 **Gearbox:** Man/Auto **Boot:** 490/1000 **Seats**: 5 **Doors:** 4/5

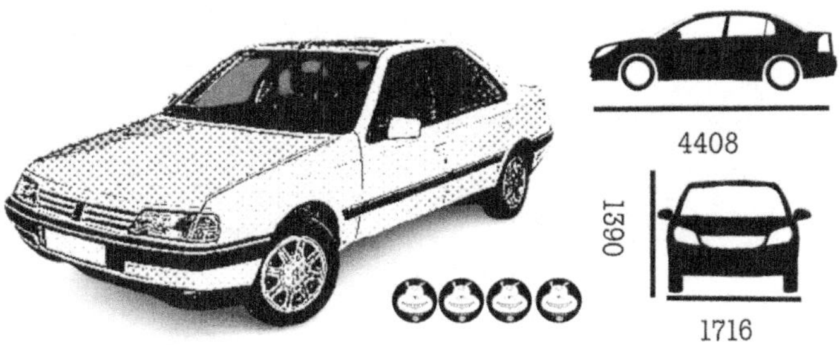

Peugeot 405 1988 - 1996

Full of '80 French style that left Cavaliers and Sierras looking very ordinaire. Nice to drive, comfortable and spacious. Massive range, small petrols struggle but the turbo diesel fabulous for the time. GTX and Executive have all the spec. 1994 is when there were upgrades for safety including drivers airbag. Worn suspension, smoky engines mostly.

Engines: 1.4 • 1.6 • 1.8 • 2.0 • 1.9D **MPG**: 30.4-47 **Gearbox:** Man/Auto **Weight**: 934 **Boot:** 470 **Seats**: 5 **Doors:** 4

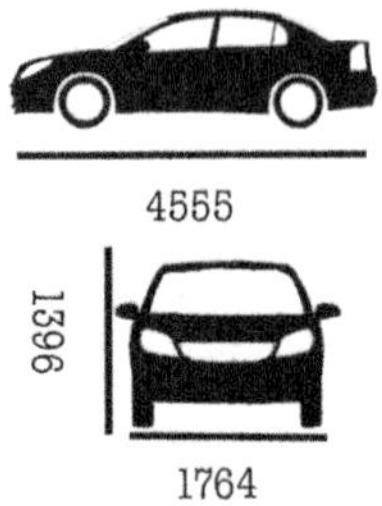

Peugeot 406 1996 - 2004

Old school Pug saloon, but all the better for that. Good to drive and simple enough to own the 1.9 Turbo is prehistoric but arguably better than most flaky modern oil burners. Otherwise the 2.0 HDi is perfect for the motorway. Revised in 2001, but our sort of Banger relic. Corroded radiators, rusty brake discs, engine ECU issues, suspension bushes.

Engines: 1.8 • 2.0 • 2.0D **CO2:** 147-197g/km **MPG**: 34.0-51.3 **Gearbox:** Man/Auto **Boot:** 526 **Seats**: 5 **Doors:** 4

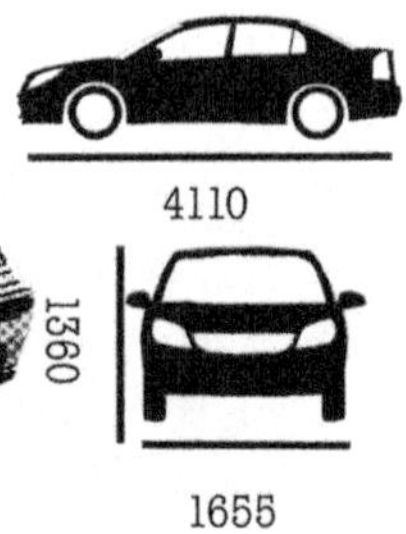

Proton Mpi 1989 - 1996

A Mitsubishi Lancer without the personality, or 4WD and wings. Deathly dull but reliable of course. Very dated '80s overall, 1.5 engine is less hard work, SE has pointless electric windows and a bodykit. GLS has power steering and a sunroof plus a radio, which is all you need. They eventually disintegrate with overwork and under servicing.

Engines: 1.3 • 1.5 **MPG**: 33.6-38.1 **Gearbox:** Man/Auto **Boot:** 400 **Weight**: 980 **Seats**: 5 **Doors:** 4/5

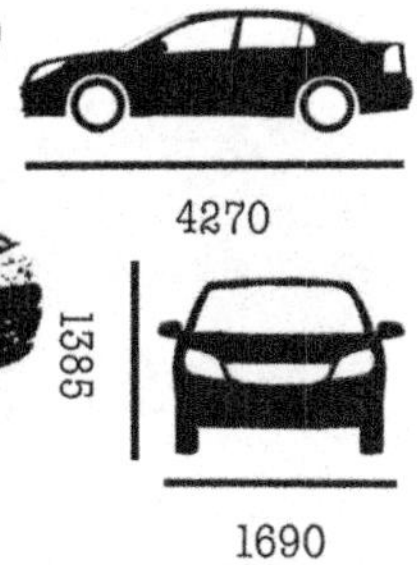

Proton Persona 1994 - 2000

Still relying on Mitsubishi, which is a jolly good thing and rather more stylish than the old MPI. Decent equipment levels EXi at the top of the spec tree, the 1.6 is the best combination of performance and economy (35mpg). Reliable, but never cherished. Just in case you wondered it became the Wira in 2000.

Engines: 1.3 • 1.5 • 1.6 2.0D **MPG**:32.1-39.8 **Gearbox:** Man/Auto **Boot:** 427-1072 **Weight:** 960 **Seats**: 5 **Doors:** 4/5

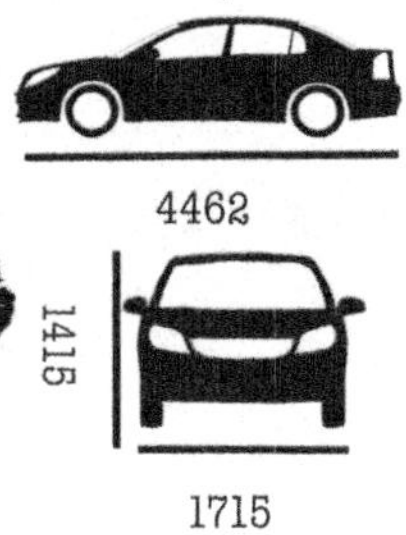

Renault 21 1986 - 1993

Spacious and smooth riding, not very stylish and a bit flakey when it comes to reliability. Diesels made sense though. Important to distinguish between Phase 3 in '92 which meant uprated specs and Phase 4 a year later but effectively this was the Savanna estate. Pretty dated by now and what finishes it off is dodgy electrics and blown engines.

Engines: 1.7 • 2.0 • 2.1D **MPG**: 30.9-46.3 **Gearbox:** Man/Auto **Weight:**1155 **Boot:** 490 **Seats**: 5 **Doors:** 4/5

Renault Laguna 1994 - 2001

Constantly updated and not a half bad hatch. Most of them have been driven into the tarmac and largely neglected. The basic diesels and petrols are the ones which have proved to be pretty dependable. Struggle to see these for sale now. Rust now a big issue, corroded heating system, failed air con, electricals always a worry, gearboxes weak, locks break.
Engines: 1.6•2.0•3.0•1.9•2.2D **CO_2:** 149-185g/km **MPG**: 24.1-51.4 **Gearbox:** Man/Auto **Boot:** 453 **Seats**: 5 **Doors:** 5

Rover 400 1990 - 1995

Proper Ronda this is and built just as well, despite K - Series shortcomings. Probably should be in the compact class, but never mind. OAP image, but comfy with side impact beams in '93 and airbag in '94 on GSi/GTi. Revised range in '92, mainly a chrome grille. Rattly camshafts, misfires, smoking diesels, there is plenty to slow them down if not cared for.
Engines: 1.4•1.6•2.0•1.8D•1.9D**MPG**: 33.1-49.8 **Gearbox:** Man/Auto **Boot:** 410 **Weight:** 1073 **Seats**: 5 **Doors:** 4/5

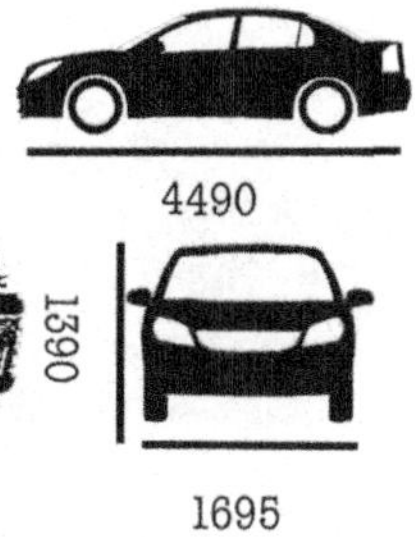

Rover 400 1995 - 1999

Still Honda related but size wise doesn't compete against a Mondeo etc. Nevertheless proper comfy, well made and there is blizzard of specifications and revisions. Executive delivers leather, CD and cruise. If you want the mostly Honda experience then the 1.6 engine with Rover trimmings in SLi spec. Otherwise, oil leaks and blown head gaskets...

Engines: 1.4 • 1.6 • 2.0 • 2.0D **MPG**: 32.1-53 **Gearbox:** Man/Auto **Boot:** 470 - 810 **Seats**: 5 **Doors:** 4/5

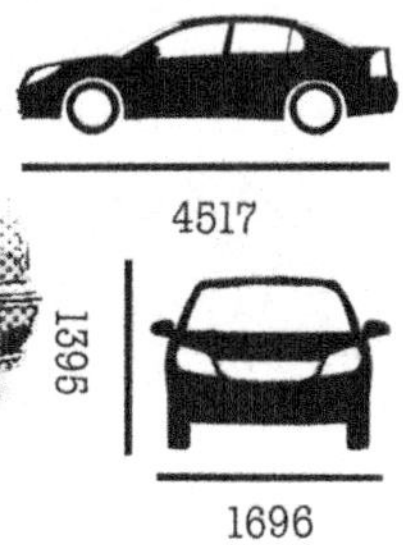

Rover 45 1999 - 2005

In a previous life a Rover 400 with that fancy grille. Still old tech rather than cutting edge. Saloon and hatches and poncy model grades, Classic, Club and Connoisseur, but all very endearing especially as a V6 and diesel. Four cylinder blown head gaskets, not so endearing. Getting rusty and parts are more of a challenge to source.

Engines: 1.4 • 1.6 • 2.0 • 2.0D**MPG**: 30.1- 52.5 **Gearbox:** Man/ Auto **Boot:** 370(hatch)/470(saloon) **Seats**: 5 **Doors**: 4/5

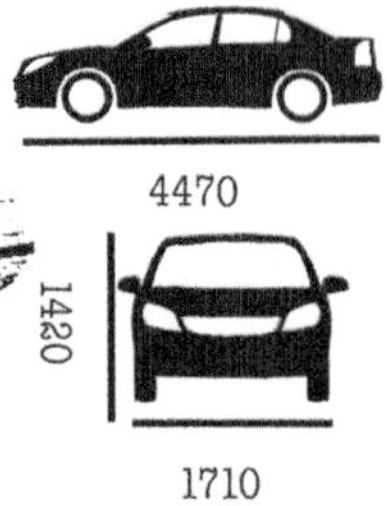

Rover Montego 1990 - 1993

Saloon that slipped into another decade and was rather outclassed. Very easy to nick, used to be the fixable Banger of choice, by the '90s just two engine options and gradually phased out, fleet order only for saloons by the end. Blown head gasket, noisy gearbox and most became terminally scruffy with most of the trim going missing or gaffer taped.
Engines: 2.0•2.0D **MPG**: 30.5-55.6 **Gearbox:** Man/Auto **Boot:** 520 **Weight**:970 **Seats**: 5 **Doors:** 5

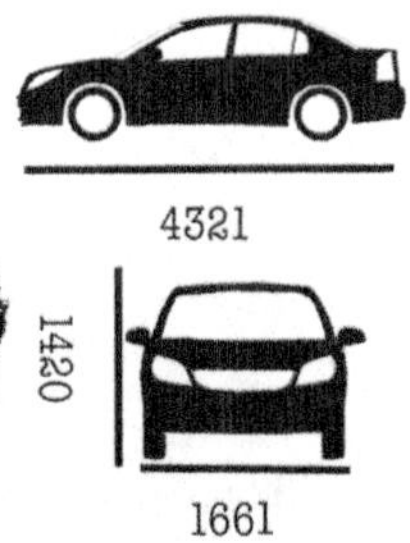

SEAT Toledo 1991- 1999

Huge hatchback that was pretty compact but as big as a Mondeo or Cavalier inside with the added benefit of VW mechanicals. It is Jetta and Corrado related but the TDI was particularly impressive. Revised in 1995 (airbag) and again in 1997, just bodywork and the alphabet soup of specifications. Oil leaks, head gaskets blown. Wear and tear.
Engines: 1.6•1.8•2.0•1.9D **MPG**: 32.1-55.4 **Gearbox:** Man/Auto **Weight:**1010 **Boot:** 549 **Seats**: 5 **Doors:** 5

SEAT Toledo 1999 - 2005

Great big booted saloon related to so many VWs we haven't got space to list them. Looked good, obviously practical and excellent engines. Cheap to fix and service which are the real plus points. Used car prices are a great leveler, but a tidy one of these, but prices should be marginal. Diesels can have ECU issues, worn clutches, minor electricals inside.

Engines: 1.8 • 2.3 • 1.9 **CO2:** 135-211g/km **MPG**: 32.1-56.5
Gearbox: Man/Auto **Boot:** 460 **Seats**: 5 **Doors:** 4

Skoda Octavia 1998 - 2005

A saloon style hatchback which remains a rather excellent way to herd your family around. Smart looking and the usually excellent line up of VW engines. Who really needs a Golf when you can have one of these? 1.9 diesels and 1.6 petrols are easy to deal with. Diesel ECU failure, water pumps, air mass sensors and ignition coils across the engine range. Electrical failures in the cockpit.

Engines: 1.4 • 1.6 • 2.0 • 1.9D • 2.0D **MPG**: 40.4-53.3
Gearbox: Man/Auto **Boot:** 560/1350 **Seats**: 5 **Doors:** 5

Subaru Impreza 1993 - 2000

Apart from the turbo nutter rally one there is the Benjy Basic saloon and hatch, the 1.6 available as a RWD as well as 4WD. LX has power steering, GL has electric windows, central locking. A good idea for towing, better mpg than a weighty SUV. Incredibly reliable and the only downside has been the cost of getting spares.

Engines: 1.6 • 1.8 • 2.0 **MPG**: 31.5-34.8 **Gearbox:** Man/Auto **Boot:** 353 **Seats**: 5 **Doors:** 4/5

Toyota Carina E 1992 - 1998

Destined to be a brilliant Banger and Brit built too. Bags of room inside and all the engines are excellent especially the tough diesels. Big news in '94 was the driver airbag, then in '96 ABS and passenger bag. Executive has it all otherwise CD will do. Smoky high mile engines, just wear and tear for suspension and other consumables.

Engines: 1.6 • 1.8 • 2.0 • 2.0D **MPG**: 32.1-48.2 **Gearbox:** Man/Auto **Weight:** 1100 **Boot:** 545 **Seats**: 5 **Doors:** 4/5

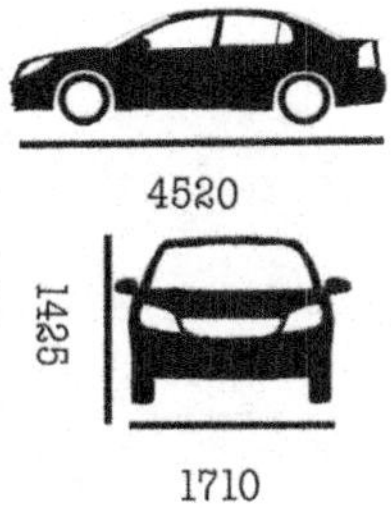

Toyota Avensis 1997 - 2003

The Brit built Carina E was fabulous and here was a saloon and Liftback based on it, but with an awful lot more safety equipment. Within a couple of years there was a diesel, the D4-D. Proper company cars that led hard lives but still managed to not break. Can still do a job. Clutch and gearbox failures over time, brake fluid leaks, just wear and tear.

Engines: 1.6 • 1.8 • 2.0 • 2.0D **CO2:** 158-181g/km **MPG**: 36.7-47.9 **Gearbox:** Man/Auto **Boot:** 510 **Seats**: 5 **Doors:** 4/5

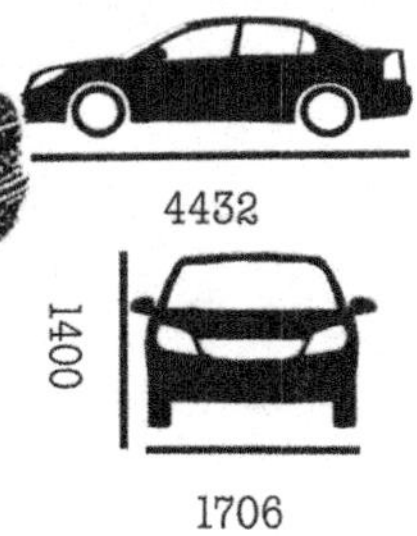

Vauxhall Cavalier 1988 - 1995

Fleet favourite from the '80s which filtered down to family duties. Practical, reliable and comfortable plus as the years went on more equipment. CD and Diplomats have it all and SRi has gas shocks and close ratio gearbox. These can be driven 24 hours a day. Oil leaks, ticking camshaft, smoky engine, worn gear linkage and suspension.

Engines: 1.4 • 1.6 • 1.8 • 2.0 • 2.5 • 1.7D **MPG**:30.8-48.9 **Gearbox:** Man/Auto **Boot:** 384 **Seats**: 5 **Doors:** 4/5

Vauxhall Vectra 1995 - 2002

Designed to thrash up and down motorways. Pretty basic stuff overall with a lot of engine and trim options. The problem is that most have been run pretty ragged and neglected, so '90s ones are properly finished. Rust obviously, leaking coolant, variously broken suspension parts, failing water pumps. Clutch remains a complicated job. **Engines:** 1.6 • 1.8 • 2.2 • 2.6 • 2.0D • 2.2D **CO2:** 161-236g/km **MPG**: 28.8-46.3 **Gearbox:** Man/Auto **Boot:** 460 **Seats**: 5 **Doors:** 4/5

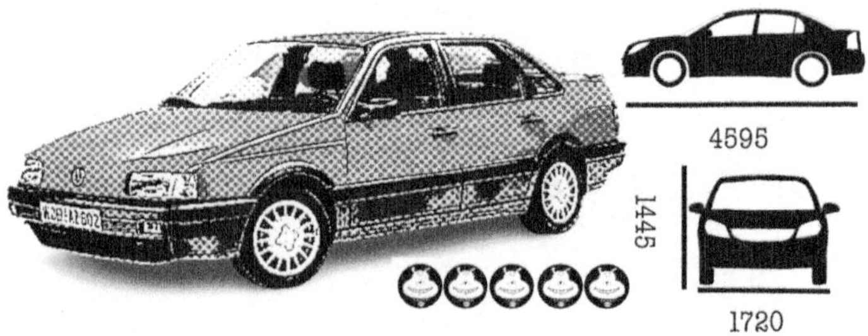

Volkswagen Passat 1988 - 1996

The last truly dull and boring one, which got better. Incredible reliability and ability to rack up massive business mileages. Superb engines better to have a 2.0 petrol or the 19TDi, not exciting even with the VR6. Late life revamp in 1994 so there's a restyle and spec upgrades with ABS on the big engines. High mile wear & tear to engine/suspension. **Engines:** 1.8 • 2.0 • 2.8 • 1.6D • 1.9D **MPG**: 28-53.3 **Gearbox:** Man/Auto **Boot:** 465 **Seats**: 5 **Doors:** 4

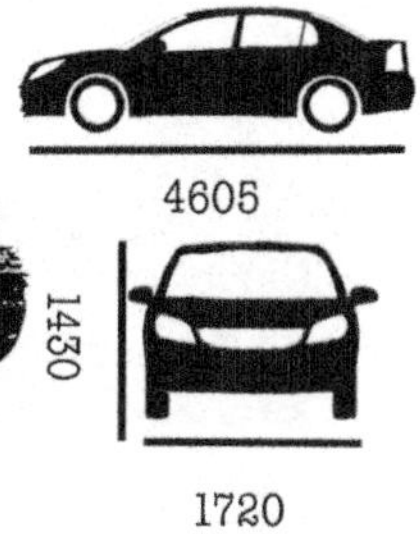

Volkswagen Passat 1997 - 2000

This is the Passat that came good. One that the company car drivers actually wanted, it actually had style. Audi A4 related but more working class. Great engines, especially the 1.9TDi, otherwise the 1.8T is the lively petrol option if you don't want 50mpg. SE and Sport and there was an air con or sunroof option. Can be driven into the ground.

Engines: 1.6 • 1.8 • 2.3 • 2.8 • 1.9D **MPG**: 26.9-53.2 **Gearbox:** Man/Auto **Boot:** 475 **Seats**: 5 **Doors:** 4

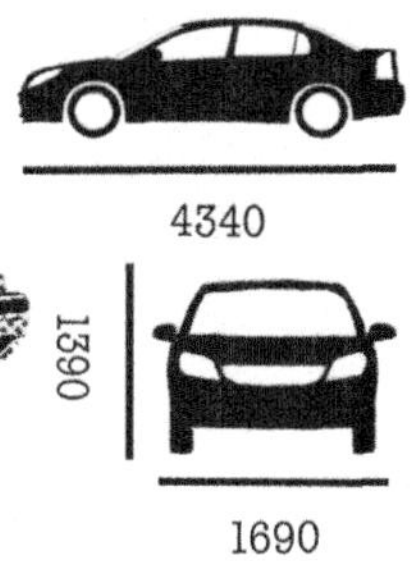

Volvo 440 1988- 1997

This Dutch built Volvo with Renault engines is better than you think despite reliability hiccups in the early days. Essentially an unremarkable family hatch that is reasonably well equipped. Full on Volvo safety in 1993 when body reinforced, airbag and pretensioners arrived. Rust plus coolant leaks, smoky engine and electrical mishaps.

Engines: 1.6 • 1.7 • 1.8 • 2.0 • 1.9D **MPG**: 32.7-36.8 **Gearbox:** Man/Auto **Boot:** 379/1027 **Seats**: 5 **Doors:** 4

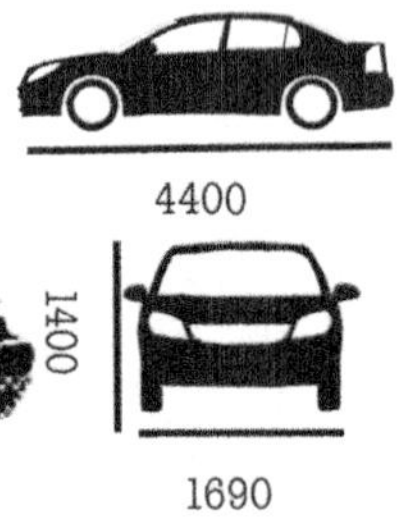

Volvo 460 1989- 1997

This is a 440 with a boot, so not sure why it merits a separate entry, but here we are. Some reckon it is more sorted than the hatch. Revised in 1993 for safety and style purposes, GLT and CD at the top of the spec tree, later ES Pack added alloys, sports suspension and metallic. Tired Renault engines can smoke, leak oil and coolant.

Engines: 1.6 • 1.7 • 1.8 • 2.0 • 1.9D **MPG**: 32.7-46.1 **Gearbox:** Man/Auto **Boot:** 453 **Seats**: 5 **Doors:** 4

WagenBangers - Matters of Estate

The hugest and most weighty chunk of the Bangerpedia is right here. May I present the estate versions of all the small family, ex-fleet and executive Bangers. This is the last gasp for one of the most popular model formats of all time. Plus we haven't even had to consider the very popular three door estate which was already dead.

Instead, big classy estates had never been more popular, usually they were Volvos. Meanwhile the Audi 100 Avant was a stylish lump that turned into the A6. The 5 Series Tourings got better as they developed from E34, into E39s. The Mercedes W124 also came out of the '80s and was a wonderfully solid example of how to do it. Shame they forgot how to do it with the later 'rusty Es'.

Here is the decade where the lifestyle estate came into its own. Although the bijou BMW 3 Series Tourings rolled up in the '80s, it was joined by the Audi 80 Avant then the A4 and the later E36 models. These were for snobs who wouldn't be seen near the back end of a common hatchback. Even Volvo offered the medium sized V40 which could cope with some relatively compact flat packs.

Then again if you do need to pick a wardrobe or two then the Citroen XM is the way to quite possibly do it. Despite the cleverness, the complexity is always a worry. The BX is a compromise and the Xantia is just a bit scary if there is a problem. That's why an equally French Peugeot 405 or 406 would be a better estate workmate on a day to day basis when there is work to do.

Although performance estates are not necessarily a part of the Bangerpedia remit, they did come our way with the RS Audis and the Subaru Imprezas, plus those quad cam Legacys. Some of them can still be Bangers including the odd V8 engined Mercedes and BMW.

There are more estate variants than you can shake a stick at. There was an immense amount of 'real estate' in this '90s collection. Good luck finding a Mitsubishi Sigma...

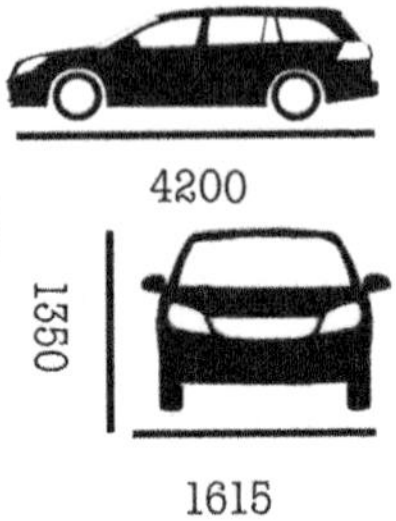

Alfa Romeo 33 Sportwagen 1983 - 1995

Alfa 33 Giardinetta was rebadged Sport Wagon in 1989. Effectively one of those lifestyle estates rather than a big lugger, but properly sporty and interestingly 1.7 QV with four wheel drive. Now had a galvanised body, revised rear suspension, all had power steering and fuel injection. there is fragile Alcantara trim. Engines need looking after.

Engines: 1.3 • 1.5 • 1.7. **MPG**: 35.8-36.7 **Gearbox:** Man **Boot:** 1200 **Weight**: 290/1005 **Seats**: 5 **Doors:** 5

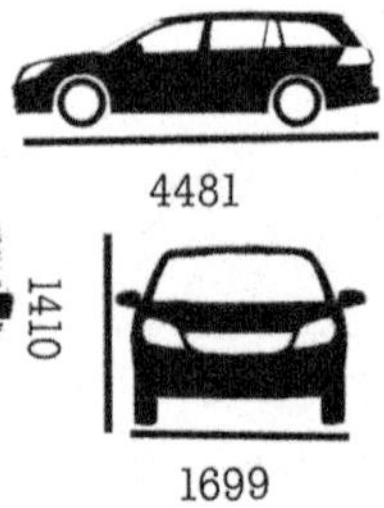

Audi 80 Avant 1991 - 1996

Lifestyle lugger, so limited space, but maximum style and built to a sky high quality level. Came with mostly zero spec and small sluggish engines. A bit tight inside and the 2.0E best entry and the V6 for smooth progress. Diesels are great for pure low level economy. Tough and just become smoky, scruffy and overworked, being galvanised helped.

Engines: 1.6 • 2.0 • 2.3 • 2.8 • 1.9D **MPG**: 31.3-44.5 **Gearbox:** Man/Auto **Boot:** 370/1200 **Doors:**5 **Seats**: 5

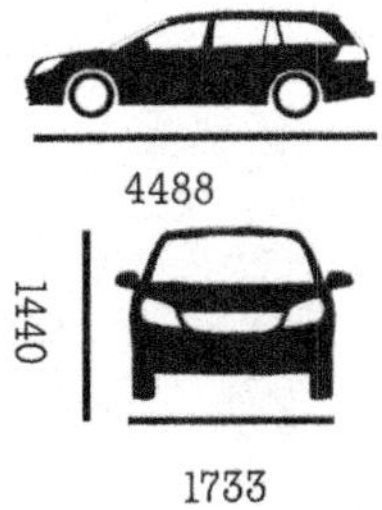

Audi A4 Avant 1996 - 2001

Default compact prestige estate from the '90s onward. From '99 a revamp tweaked the suspension and steering for the better. Good size load area, that is wide and flat, nothing innovative, just carpeted quality. Hardly huge, it has oddments space but won't take a wardrobe. Wear and tear issues of brakes and suspension, otherwise it is service history. **Engines:** 1.6 • 1.8 • 2.4 • 2.6 • 1.9D • 2.5D **MPG**: 23-53 **Gearbox:** Man/Auto **Boot:** 390/1430 **Seats**: 5 **Doors:** 5

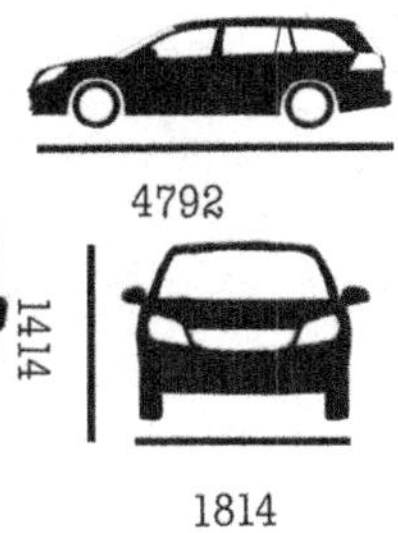

Audi 100 Avant 1991 - 1994

Slippery galvanised estate car legend makes it into the '90s under its own power. Safe, comfortable and refined, with some great engines, from the TDI diesels to the smooth V6. There was even optional 6 speed gearboxes and Quattro all wheel drive. So can get complicated, smoky engines, noisy gearboxes and bearings as the miles mount up.

Engines: 2.0 • 2.2 • 2.3 • 2.6 • 2.8 • 2.5D **MPG**: 26.1-44.1
Gearbox: Man/Auto **Boot:** 390/1310 **Seats**: 5

Audi A6 Avant 1998 - 2004

Executive transportation. Even Audi described it as a 5 door coupe with the usual ton of engine and drivetrain options. Avoid the Tiptronic clutchless manual on the grounds of complication. Fuel pumps and air flow meters can fail, wheel bearings, plus with old age electric windows and central locking. **Engines**: 1.8•2.0•2.4•2.7•3.0•4.2•1.9 •2.5D **MPG**: 21.7-49.6 **Gearbox:** Man/Auto **Boot:** 455/1591 **Seats**: 5 **Doors:** 5

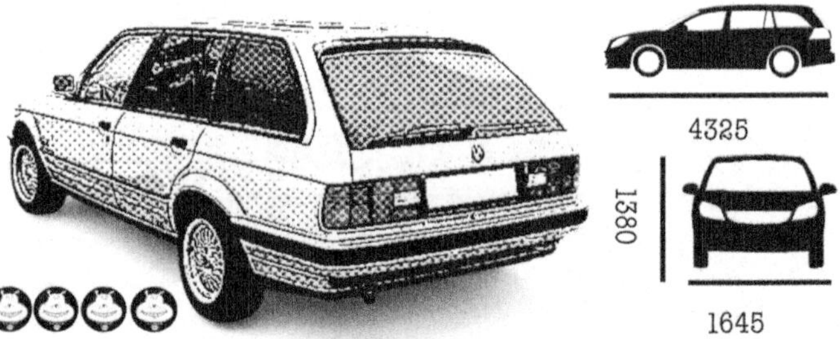

BMW 3 Series Touring 1988 - 1993

Ultimate lifestyle estate thrives in a new decade along with other variations, but deserves its place here as one of the best pre modern Bangers. Great 6 cylinder engines, the fours will plod on though. Late model uprated specs SEs with cross spokes and spoilers. Killed by rot, blown head gaskets, worn out suspension and just general neglect in the old days. **Engines:** 1.6•1.8•2.0•2.5 **MPG**: 30.1-35.3 **Gearbox:** Man/Auto **Boot:** 371/1125 **Seats**: 5 **Weight:** 1190

BMW 3 Series Touring 1995 - 1999

Not quite as E30 solid but probably slicker and certainly fine to own and drive, Everyone wanted one. Usual slant of good fours, great sixes and arrival of the impressive turbo diesels. Again the 6-cylinder diesel was astounding. Revamped in '96 a bit, Sports, SE and Lux packs. Pretty reliable, but fluid leaks, iffy trim fragile, some electrics go 'bong'.

Engines: 1.6 • 1.8 • 2.0 • 2.5 • 2.8 • 1.8D • 2.5D **MPG**: 31.0-50.4 **Gearbox:** Man/Auto **Boot:** 370/1320 **Seats**: 5 **Doors:** 5

BMW 3 series Touring 1999 - 2005

Handsome 3 series with a great range of engines both petrol and diesel. Averagely spacious as it isn't a real estate. Has floor hooks and straps and storage under the floor and the clever split tailgate glass for convenience. Standard SE spec is good. Power loss and smoke, diesel issues, broken air mass meters and cracked inlet manifolds. Worn balljoints.

Engines: 1.8 • 2.0 • 2.2 • 2.5 • 3.0 • 1.8D • 2.0D • 3.0 **MPG**: 31.0-50.4 **Gearbox:** Man/Auto **Boot:** 435/1345 **Seats**: 5

BMW 5 Series Touring 1992 - 1997

Models goes up a few gears in comfort, quality and sheer presence, the premier Executouring banger of the time. 4WD X model in 1992. V8s are great, but a big 6 is the decent compromise. 518 can't cope, 530i can. Plenty to go wrong if not cared for, electronics, suspension bushes injectors on diesels, whilst there is smoke and gearbox faults.
Engines: 1.8•2.0•2.5•3.0•3.5•4.0•2.5D **MPG**: 22.3-39.6 **Gearbox:** Man/Auto **Boot:** 460/910 **Seats**: 5 **Doors:** 5

BMW 5 Series Touring 1997 - 2004

The last of the fixable BMWs and one of the greatest BMWs ever. More practical than before, can tow 2000kg, has self-leveling suspension, a load net and the split fold rear seats lie flat. The diesels are great especially the bigger sixes and the petrol sixes will last forever, if not longer. Electronics on old ones, suspension bushes. Diesels may need new injectors.
Engines: 2.0•2.2•2.5•3.0•3.5•4.0•2.5D•3.0D **MPG**: 23.0-42.2 **Gearbox:** Man/Auto **Boot:** 408/1521 **Seats**: 5 **Doors:** 5

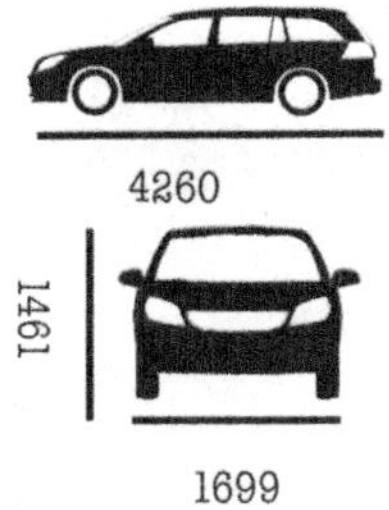

Citroen ZX Estate 1994 - 1998

Hard working estate called Break abroad, it just gets on with family job and with the unburstable basic diesels that will run on chip fat make it a cockroach used buy. Rear seats slide forward on most models. Electrics may play up. Worn out suspension, starting and running problems due to blocked injection systems.

Engines: 1.4 • 1.6 • 1.8 • 2.0 • 1.9D **MPG**: 33.3-49.1 **Gearbox:** Man/Auto **Boot:** 487/1580 **Seats**: 5 **Doors:** 5

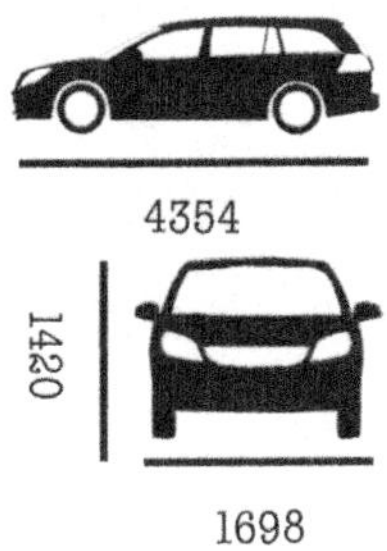

Citroen Xsara Estate 1997 - 2001

A sexed up ZX which was always reliable enough and pretty frugal on the old fuel whether you go for a 1.4 or 2.0 HDi. Load area is flat and wide. Does not seem to have many reliability downsides provided it has not been run into the ground. ECU issues, some late '90s diesel engine issues can kill the engine. Diesel clutches, also tatty bodywork.

Engines: 1.4 • 1.6 • 2.0 • 1.9D • 2.0D **MPG**: 40.9-54.3
Gearbox: Man/Auto **Boot:** 512/1512 **Seats**: 5 **Doors:** 5

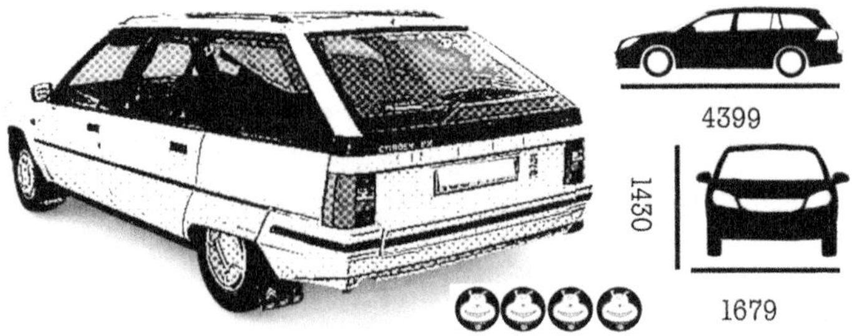

Citroen BX Estate 1983 - 1994

The Phase II models from 1990 still had a colossal amount of room with the rear seat folded. Excitingly rare 4×4. Super comfy, practical and used to be cheap because of the complexities. The diesels are old school but great. Plenty to go wrong, suspension and steering, camshaft wear, snapped clutch cables, general electric upsets. '80s icon

Engines: 1.6 • 1.9 • 1.7D • 1.9D **MPG**: 33.8-48.2 **Gearbox:** Man/Auto **Boot:** 512/1803 **Seats**: 5 **Doors:** 5

Citroen Xantia Estate 1995 - 2001

Arrival of the Euro Badged Break model delivered loads of room. Hydractive suspension is great for comfort and can be terrible when its a FrancoBanger. Otherwise lovely to drive, great engines from V6 to the turbo diesels. Phase 2s in '98 which are stronger and safer. Apart from the hydraulic related issues, ECUs, diesel injectors.

Engines: 1.6 • 1.8 • 2.0 • 3.0 • 1.9D • 2.1D **MPG**:24.4-44.7 **Gearbox:** Man/Auto **Boot:** 610/1690 **Seats**: 5 **Doors:** 5

Citroen XM Estate 1992 - 2000

Massive amounts of room, spectacular use of comfy self-level ling suspension.. The manuals are bit clunky, and 2.1 diesel agricultural but what a wonderful way to travel. Excellently space age interior and dash. Requires bravery to own with slipping auto gearboxes, broken clutches on the manuals and all sorts of fun with Hydractive suspension.

Engines: 2.0 • 3.0 • 2.1D • 2.5D **MPG**: 24.4-41.9 **Gearbox:** Man/Auto **Boot:** 480/1755 **Seats**: 5 **Doors:** 5

Daewoo Nubira Estate 1997 - 2002

Estate car with a long flat loading bay is all any buyer wants. It doesn't deliver much else except an unrefined 1.6 engine, but it is reliable, the fixtures and fittings are good the SX includes air con and alloys. Quite likely to be neglected and worn out, but still working. Nothing to look for, just wear and tear items like brakes and suspension.

Engines: 1.6 • 2.0 **MPG**: 32.8-36.2 **Gearbox:** Man/Auto **Boot:** 549/1838 **Seats**: 5 **Doors:** 5

Fiat Tempra SW 1991 - 1996

Perfect family estate, but few sold as the dullness put most off. Cheap used and therefore abused. Actually good to drive. Got safer from '93 with a stronger body, side impact bars and airbag. S, SX and SLX, avoid basic ones, but SX has digital instruments that may have gone AWOL. Worn valve guides, iffy gearboxes and CVT.

Engines: 1.4 • 1.6 • 1.8 • 2.0 • 1.9D **MPG**: 33.7-47 **Gearbox:** Man/Auto **Boot:** 500/1550 **Seats/Doors**: 5

Fiat Marea Weekend 1997 - 2003

Stylish thing with a huge boot that got huger with the rear seats folded. It is nice to drive, but the fact it is a Fiat meant no one took it seriously except those who actually ran them and most are very happy. If any are around, should cost buttons. Chocolate ones. Suspension drop links, sunroof leaks, instrument panel failures, needs red master key.

Engines: 2.0 • 1.9D **CO2:** 149-234g/km **MPG**: 28.8-50.4 **Gearbox:** Man/Auto **Boot:** 501/1549 **Seats**: 5 **Doors:** 5

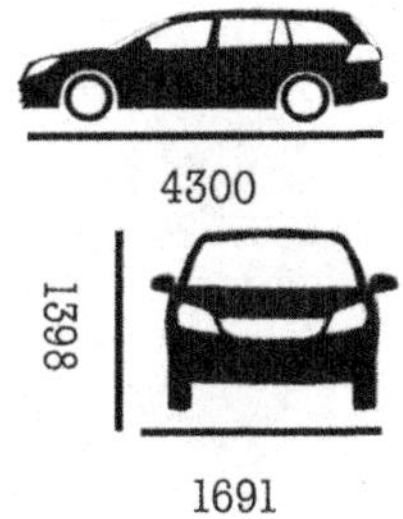

Ford Escort Estate 1990 - 1995

Regarded as a cynical collection of new parts that produced a dreary and underwhelming old thing. Still sold in big numbers, estate always made sense for junior reps. Got revamped in '92 with restyle and new Zeta engines. Ghia estates have roof rack. Electricals never great, ECUs and windows, noisy camshaft, tappets, snapped cam belts.

Engines: 1.3 • 1.4 • 1.6 • 1.8 • 1.8D **MPG**: 25.6-53 **Gearbox:** Man/Auto **Boot:** 360/1100 **Seats**: 5 **Doors:** 5

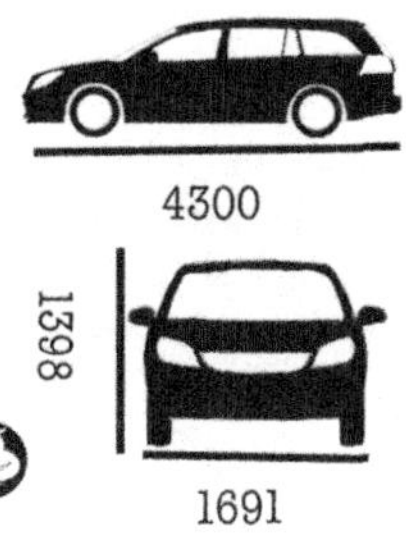

Ford Escort Estate 1995 - 2001

Mark 7 lived on as a special edition in the Focus era. Flight and Finesse with 1.6 petrol and 1.8 Turbo Diesel. Rust, neglect and abuse killed most of them. Looks and drives prehistoric, but a solid used buy. Blown head gaskets, bodywork rust serious, rusty fusebox, can be fatal. Minor electricals bring them to a halt as well.

Engines: 1.6 • 1.8D **MPG**: 42.7-44.2 **Gearbox:** Man/Auto **Boot:** 380/1130 **Seats**: 5 **Doors:** 5

Ford Focus Estate 1998 - 2004

Long wheelbase means there is loads of luggage space, plus everyone loves a Focus as they are great to drive and easy to own. Low loading lip and well packaged suspension TDCI diesels are great Zetec spec and the petrols are all pretty reliable. Getting old a patchy now though and starting to rust. Otherwise it is just wear and tear to the suspension.

Engines: 1.4 • 1.6 • 1.8 • 2.0 • 1.8D **CO2:** 140-188g/km **MPG**: 30.0-53.3 **Gearbox:** Man/Auto **Boot:** 521/1580 **Seats**: 5

Ford Sierra Estate 1982- 1993

Jellymould extension had lots of room and was certainly wildly popular as a hand me down from company fleets to families. It is like working on an old Cortina which makes it one the better recent period Bangers. Loadsa specs and specials, but ok from the LX upwards. Wear and tear which is all fixable apart from terminal bodywork.

Engines: 1.6 • 1.8 • 2.0 • 2.9 • 1.8D **MPG**: 26.1-38.7 **Gearbox:** Man/Auto **Boot:** 427/1958 **Seats**: 5 **Doors:** 5

Ford Mondeo Estate 1993 - 1996

Massively better than the Sierra. Not a looker, but great to drive, practical, comfortable. Like the Sapphire needs to be GLX and above, whilst the Ghia X has knobs on like cruise, leather and a CD. Four wheel drive again and there's a sporty Si with revised suspension. Can have a sticky clutch and worn suspension. Mostly suffers from serious wear.

Engines: 1.6 • 1.8 • 2.0 • 2.5 • 1.8D **MPG**: 30.6-46.3 **Gearbox:** Man/Auto **Boot:** 540/1515 **Seats**: 5 **Doors:** 5

Ford Mondeo Estate 1996 - 2000

Enough of an update to warrant Mondeo Mark II status with a new face plus uprated suspension, engines, gearbox and the interior. Good standard spec above the Aspen and Verona, so air CD system and Quicklear windscreen. More to go wrong, but mostly it doesn't if looked after, but multi owners and mega mileages eventually wore them out.

Engines: 1.6 • 1.8 • 2.0 • 2.5 • 1.8D **MPG**: 30.6-46.3 **Gearbox:** Man/Auto **Boot:** 540/1515 **Seats**: 5 **Doors:** 5

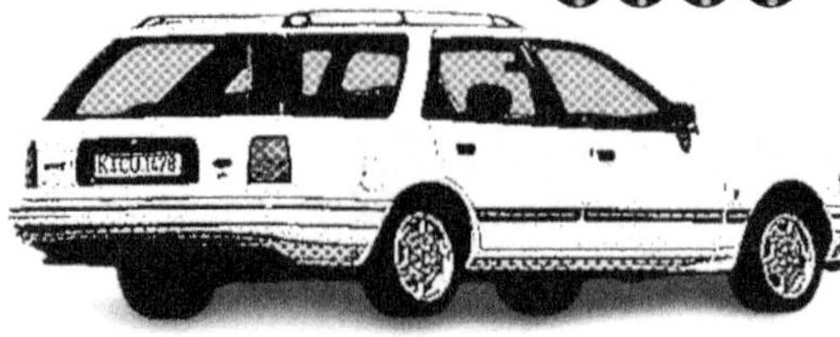

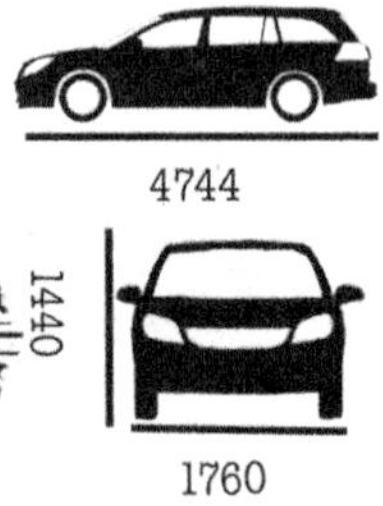

Ford Granada Estate 1992 - 1994

From 1990 revamped Granny, pushed upmarket 9estate a couple of years later) especially in Scorpio trim (above Ghia and Ghia X) and with the full fat 2.9 24V. The turbo diesel is a bit of slug, 2.0 petrol best all round, lots of room, was easy and cheap to fix at the time. Loads of smoke, worn cranks, wheel bearings and soggy suspension.

Engines: 2.0 • 2.9 • 2.5D **MPG**: 24.8-35.6 **Gearbox:** Man/Auto **Boot:** 550/1600 **Seats**: 5 **Doors:** 5

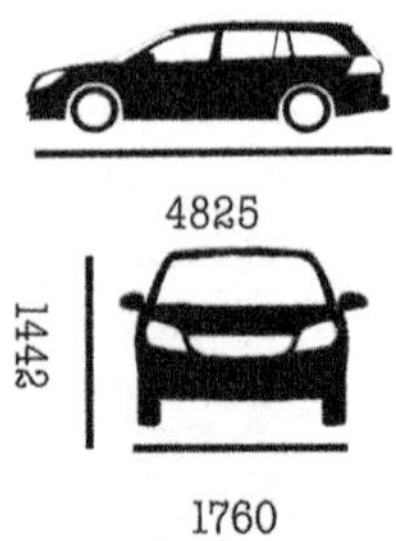

Ford Scorpio Estate 1996 - 1999

Here is the Guppy one, designed to be different and last of the Execufords. Six engines on offer, and loads of specs the Executive was the basic one, then Ghia, Ghia X and Ultima. All had useful 60/40 folding seat, metallic and music system. Late life minor restyle in '98. Unloved at the time, had minor electricals e.g. windows, otherwise engines and suspension.

Engines: 2.0 • 2.3 • 2.9 • 2.5D **MPG**: 23.9-32.8 **Gearbox:** Man/ Auto **Boot:** 550/1600 **Seats**: 5 **Doors:** 4/5

Honda Civic Shuttle 1984 - 1991

So called 'tall' vehicle which isn't quite an MPV more of a breeze block estate. There was a 4WD for a bit. Didn't catch on, not really space shuttley enough, Civic reliable though. Otherwise only around briefly in the 1990s before the name returned on a proper people shifter. This is just a weird Civic incarnation which now seems attractively oddball.

Engines: 1.4 • 1.5 • 1.6 **MPG**: 37.6-39.9 **Gearbox:** Man
Weight: 1050 **Boot:** 328/648 **Seats**: 5 **Doors:** 5

Honda Civic Aerodeck 1995 - 2001

Generation 6 built in Britain and Rover 400 related, based on Honda Domani, it delivers a reasonable amount of space, with a deep and wide boot, but some stats suggest just 835 litres with seats folded. Even if the interior is compromised it is just relentlessly capable. That's a good thing. No real issues all, rust is probably more of an issue which can kill it.

Engines: 1.5 • 1.6 • 1.8 **MPG**: 32.1-47.5 **Gearbox:** Man/Auto
Boot: 415/1312 **Seats**: 5 **Doors:** 5

Honda Accord Aerodeck 1989 - 1993

Shockingly huge and overlooked model which is unbreakable and comfortably dull. Not a great deal to say about these apart from the obvious, which is the cost of repairing one these days. The uninjected 2.0 is probably worth avoiding, otherwise there is an outside chance of a blown head gasket, worn gearbox and leaky power steering.

Engines: 2.0 • 2.2 **MPG**: 29.1-32.7 **Gearbox:** Man/Auto **Boot:** 980/1830 **Turning**: 10.8 **Seats**: 5 **Doors:** 5

Honda Accord Aerodeck 1994 - 1997

American built model not around for very long. Got a chrome grille in '96 to make it look more Brit, probably. Otherwise available as a 2.0 LS and 2.2 ES. Not much happened apart from the grille. These are tough. solidly sensible and fairly comfortable. No faults at all, Just rot and neglect at this stage of life.

Engines: 2.0 • 2.2 **MPG**: 31.4-32.4 **Gearbox:** Man/Auto **Boot:** 430/739 **Seats**: 5 **Doors:** 5

Lada Riva Estate 1985 - 1996

Cheap and cheerless old Fiat that was shockingly popular on account of its dirtcheapness. Not much to recommend it, especially a few neglectful owners down the line. Range revised in 1991 into Es and Ls, '92 revamp as 1.5L gets colour coded grilles, remote adjustable door mirrors and a stereo cassette. Cats from '93. Noisy diff, knocking big ends.

Engines: 1.3 • 1.5 • 1.6 **MPG**: 29.7-31.5 **Gearbox:** Man
Boot: 500/1000 **Fuel tank**: 46.8 L **Seats**: 5 **Doors:** 5

Mazda 323 Estate 1989 - 1994

Forgotten, conventional small format estate which looked like a Ford Escort and was just as dull. Injected engines from 1991, modded suspension and an optional automatic with the 1.6 engine. Side impact beams from '93. Overall incredibly reliable and hard to fault unless neglected or the servicing regarded as optional.

Engines: 1.3 • 1.6 • 1.8 **MPG**: 33.8-38.3 **Gearbox:** Man/Auto
Boot: 378/1600 **Seats**: 5 **Doors:** 5

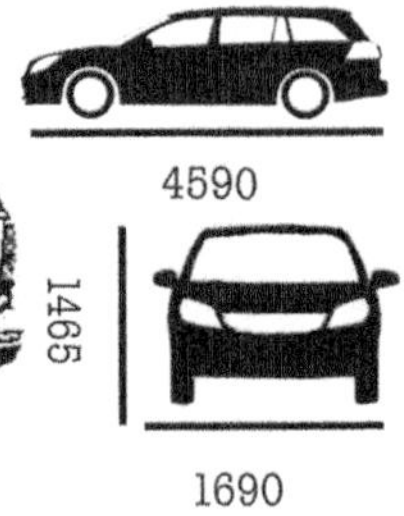

Mazda 626 Estate 1987 - 1991

Very conventional small station wagon for the American market. Took the doors from the hatch and positioned the rear lights to make a large tailgate that opened to the bumper. Large flat loading area. Super useful and practical especially as all the mechanicals are ultra reliable Mazzers. Not cool, just useful, most run deep into the ground.
Engines: 1.8•2.0 **MPG**: 25 **Gearbox:** Man/Auto **Boot:** 430/1200 **Weight**: 1200 **Seats**: 5 **Doors:** 4/5

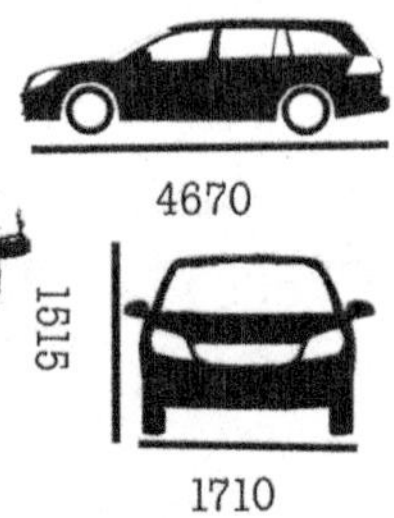

Mazda 626 Estate 1992 - 1997

Well equipped and solidly put together middle order estate car which is incredibly reliable and hard to fault. GLX delivers most of the spec you need including electric door mirrors whilst an Executive adds velour, plus a touch of walnut and cruise. Not much goes wrong, if you can find one after it's done a seven figure mileage as a mundane minicab.
Engines: 1.8•2.0•2.5•2.0D **MPG**: 29.3-41.1. **Gearbox:** Man/Auto **Boot:** 484/1677 **Weight**: 1330 **Seats**: 5

Mazda 626 Estate 1997 - 2002

Very much an anti-climax of a car, does everything averagely well, but what it excels at is not breaking down. Top notch SE delivers alloys, cruise and some pointless wood trim. The restyle in 1999 was meant to help but it was still dreary. It's an old school minicab because it is roomy enough and needs just basic servicing to keep on working,

Engines: 1.8•2.0•2.0D **MPG**: 34.9-54.3 **Gearbox:** Man/Auto **Boot:** 484/1677 **Weight**: 1330 **Seats**: 5

Mercedes C-Class Estate 1996 - 2001

Rear seats split/fold and roof bars are standard, but that was all you needed back then. The 1997 revamp was considerable with new engines and silly subtle styling things like dark tail lights, if that was your bag. The diesels are great of course. Still rusted. Indeed rusty suspension parts, wipers cost a bomb to fix, engine oil leaks.

Engines: 1.8•2.0•2.3•2.4•2.8•2.2D•2.5D **MPG**: 26.6-45.6 **Gearbox:** Man/Auto **Boot:** 465/1510

Mercedes E-Class Estate 1984 - 1995

Epically reliable, comfortable and nothing Merc has ever made since is as solid. Colossal number of engines. Mercedes added more safety over the years so an airbag in '92 and a passenger one in '94. Rear facing kiddie seats option. Massive mileages and neglect take their toll, the expensive bits are automatic gearboxes and rear axle. Magnificent.

Engines: 2.0 • 2.2 • 2.6 • 2.8 • 3.0 • 3.2 • 2.5D • 3.0D
MPG: 19.6-38.9 **Gearbox:** Man/Auto **Boot:** 530/1770

Mercedes E-Class Estate 1996 - 2003

Es were good, but these were disappointingly rusty. Usual diesels and petrols and substantially revamped in 1999, so better to buy one of those if you must. The common rail diesels are good, eventually illegal and the petrols reliable. Oil leaks, rust to brake lines, suspension bushes, engine mountings. More Mercedes branded rust.

Engines: 2.0 • 2.4 • 2.8 • 3.2 • 4.3 • 2.1D • 2.7D • 3.2D **MPG**: 25.0-45.6 **Gearbox:** Man/Auto **Boot:** 538/1974 **Seats**: 5

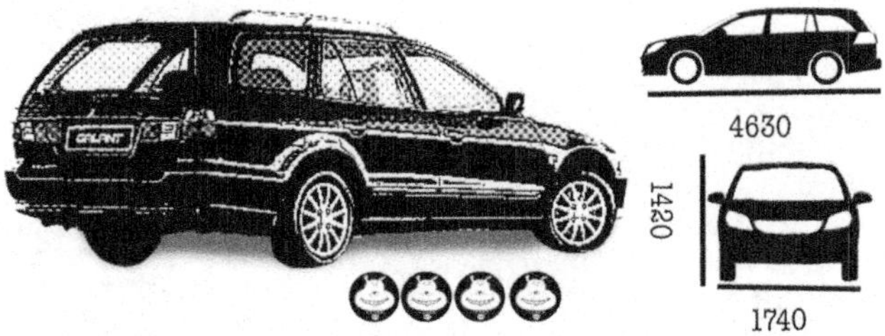

Mitsubishi Galant Estate 1997 - 2003

When a buyer needs plenty of space and has miles to cover this is perfect. Contained a GDI engine, a gas direct injection unit which was jolly efficient even if Euro fuel wasn't always pure enough for it. Not a big hit, few around, classy and solid, proper alternative to one of those German estate cars. Check gearbox condition, brakes and suspension.

Engines: 2.0•2.4•2.4 **CO2:** 203-207g/km **MPG**: 29.7-33.2
Gearbox: Man/Auto **Boot:** 419/1399 **Seats**: 5 **Doors:** 5

Mitsubishi Sigma Estate 1991 - 1995

Japanese 5 series and largely forgotten. Had trick four wheel steering, electronically controlled suspension so super clever and complicated. Leather and cruise. Revised a bit in '93. Got high intensity headlamps and uprated trim a year later and that was it. Highly reliable but needed pukka looking after. No image, meant no one cared.

Engines: 3.0 **MPG**: 24.7-25.6 **Gearbox:** Man/Auto
Boot: 447 /737 **Towing**: 1800 **Seats**: 5 **Doors:** 5

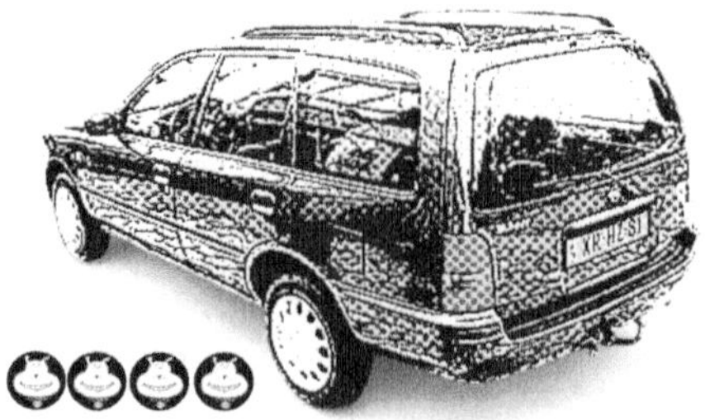

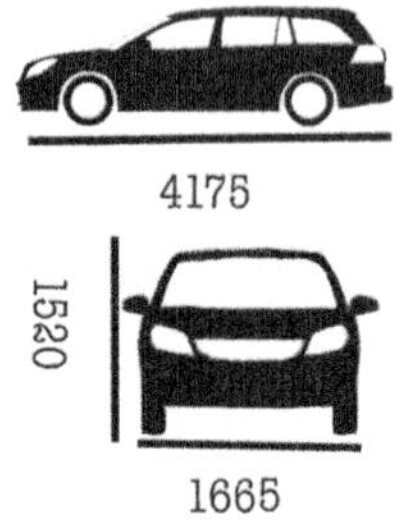

Nissan Sunny Estate 1991- 1995

Legendary old lag that made it into the '90s in good health. Always great value, it was mostly cashed strapped families and OAPs that invested heavily in this dull but worthy wagon. '93 saw injection engines and side impact beams through the range. Old age means oil leaks, chirping camshaft and collapsing trim.

Engines: 1.6 **MPG**: 39.5 **Gearbox:** Man **Boot:** 460/1600 **Max Load**: 494 **Seats**: 5 **Doors:** 5

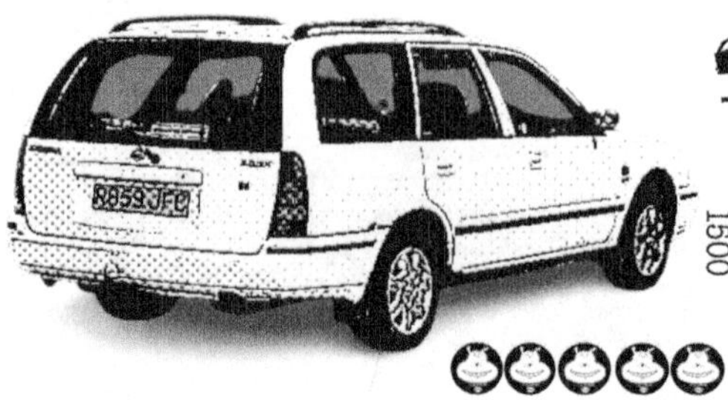

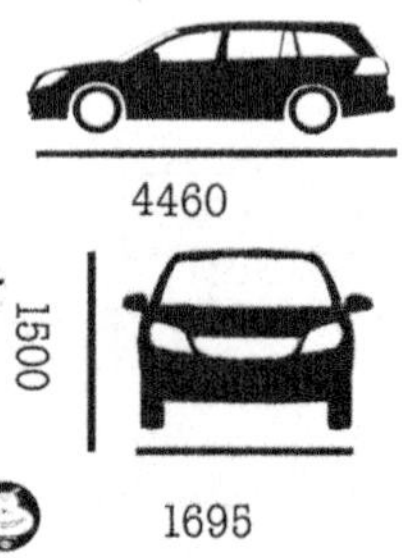

Nissan Primera Estate 1991 - 1996

It could not be less spectacular, but that is why it's spot on. A fleet car legend which got on with the job, by not breaking down and lugging loads of stuff without incident.. Series 2 in '93 with side impact beams and more equipment. SE has leather, alloys, wood and CD. Diesel not that frugal, whereas the 1.6 was. Survivor of extreme wear and tear and neglect.

Engines: 1.6 • 2.0 • 2.0D **MPG**: 33.4-39.1 **Gearbox:** Man/Auto **Weight**: 1229 **Boot:** 460/1700 **Seats**: 5 **Doors:** 5

Nissan Primera Estate 1998 - 2002

Blandness in estate format with lots of standard kit of course including airbags, ABS brakes and climate control on all but the basic example. The engines are great, go for the 1.8 or 2.0 petrol or the turbo diesel. It will only be consumables like tyres and exhausts that may need replacing. Older examples are starting to rot, suspension wear, air con busted, old alloys corrode.

Engines: 1.6 • 1.8 • 2.0 • 2.0D **CO2:** 179-195g/km **MPG**: 34.8-43.5 **Gearbox:** Man/Auto **Boot:** 430/1651 **Seats**: 5 **Doors:** 5

Peugeot 306 SW 1997 - 2002

Partly Citroen ZX with less dreary bodywork. Connected to the basic 1.9 and the later Hdi diesels it was brilliant. These days a 1.6 petrol is all any buyers needs to keep on motoring, they won't be cool though as the air con pipes on the higher spec ones rot away. Electrics are the let down. Worn suspension, driveshafts, rust, mostly worn out.

Engines: 1.4 • 1.6 • 1.8 • 1.9D • 2.0D **CO2:** 141-198g/km **MPG**: 33.6-54.3 **Gearbox:** Man/Auto **Boot:** 442/1512 **Seats**: 5

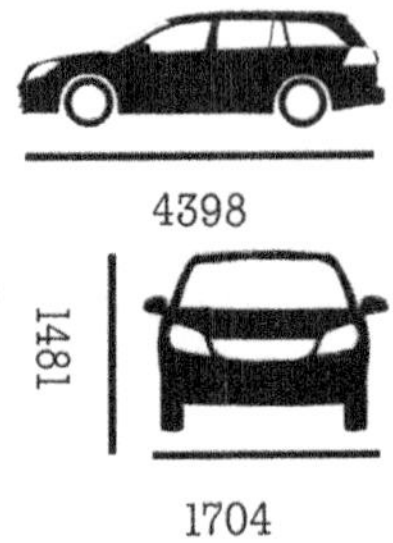

Peugeot 405 Estate 1988 - 1996

Full of '80 French style in ultra useful Famailie format.. Nice to drive, comfortable and spacious. Massive range, small petrols struggle but the turbo diesel fabulous for the time. GTX and Executive have all the spec. 1994 is when there were upgrades for safety including drivers airbag. Worn suspension, smoky engines mostly.

Engines: 1.6• 1.8•2.0•1.9D **MPG**: 30.4-47 **Gearbox:** Man/ Auto **Weight**: 934 **Boot:** 425/1640 **Seats**: 5 **Doors:** 5

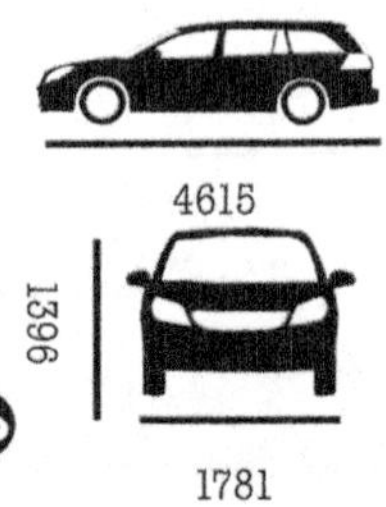

Peugeot 406 Estate 1996 - 2004

It is usefully long with a low sill with flipping and folding the seats leaves a large vanlike amount of space. 1.9 Turbo is prehistoric but arguably better than most flaky modern oil burners. Otherwise the 2.0 HDi is perfect for the motorway. Revised in 2001,our sort of banger relic. Corroded radiators, rusty brake discs, engine engine ECUs, suspension bushes.

Engines: 1.8•2.0•2.0D **CO2:** 147-197g/km **MPG**: 34.0-51.3 **Gearbox:** Man/Auto **Boot:** 526/1741 **Seats**: 5 **Doors:** 5

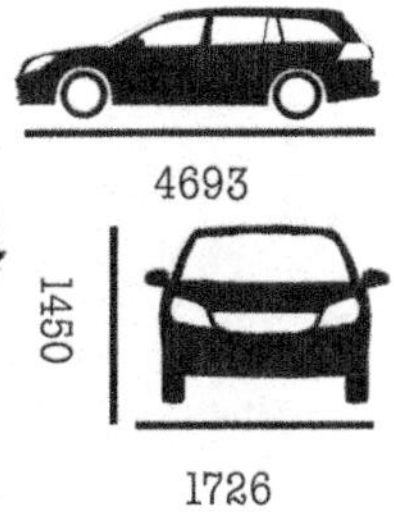

Renault 21 Savanna 1986 - 1993

Spacious and smooth riding, not very stylish and a bit flakey when it comes to reliability. Diesels made sense though. Important to distinguish between Phase 3 in '92 which meant uprated specs and Phase 4 a year later but effectively this was the popular 7 seat Savanna estate. Pretty dated by now and what kills it is dodgy electrics and blown engines.

Engines: 1.7 • 2.0 • 2.1D **MPG**: 30.9-46.3 **Gearbox:** Man/Auto Weight:1155 **Boot:** 674/1200 **Seats**: 5/7 **Doors:** 5

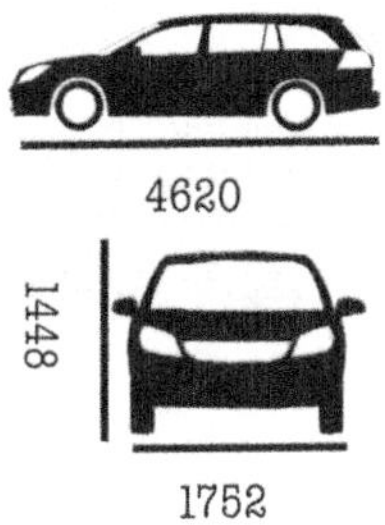

Renault Laguna Estate 1996 - 2001

Reasonable old lugger. Most of them have been driven into the tarmac and largely neglected. The basic diesels and petrols are the ones which have proved to be pretty dependable. Struggle to see these for sale now. Rust now a big issue, corroded heating system, failed air con, electricals always a worry, gearboxes weak, locks break.

Engines: 1.6 • 2.0 • 1.9 • 2.2D **MPG**: 33.1-51.4 **Gearbox:** Man/ Auto **Boot:** 520/1782 **Seats**: 5 **Doors:** 5

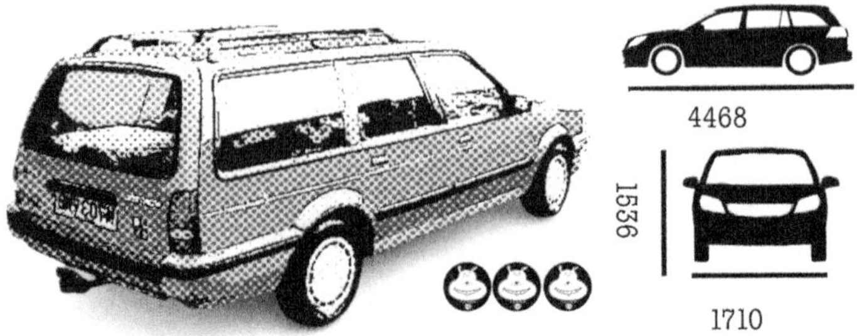

Rover Montego Estate 1990 - 1993

Surprisingly classy version, especially in Countryman spec. Rear facing child seats, so family friendly Very easy to nick, used to be the fixable Banger of choice, by the '90s just two engine options and gradually phased out. Blown head gasket, noisy gearbox and most became terminally scruffy with most of the trim going missing or gaffer taped in place.

Engines: 2.0 • 2.0D **MPG**: 30.5-55.6 **Gearbox:** Man/Auto **Boot:** 509/862 **Weight**:970 **Seats**: 5/7 **Doors:** 5

Rover 400 Tourer 1994 - 1999

Stylish old thing and the Honda bodes well, but it isn't hyper quality as the interior gets pretty shabby with neglect. Probably best to stick with the 1.6 Honda engine rather than the Rover VVC, but the diesel does an economy job well enough. Rattly camshafts, misfires, smoking diesels, there is plenty to slow them down if not cared for.

Engines: 1.6 • 1.8 • 1.8D **MPG**: 33.1-49.8 **Gearbox:** Man/Auto **Boot:** 430/1410 **Weight:** 1125 **Doors:** 5

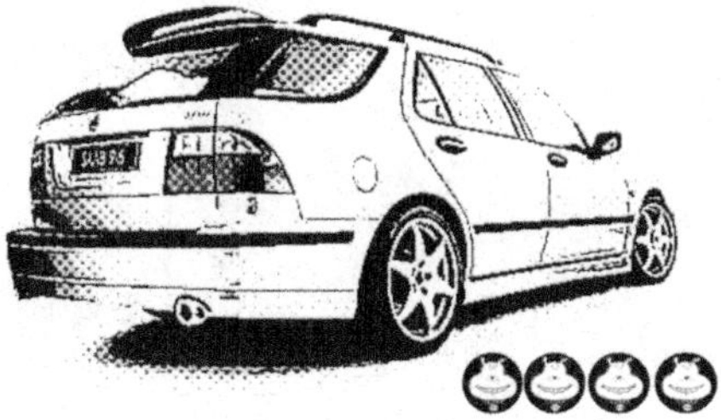

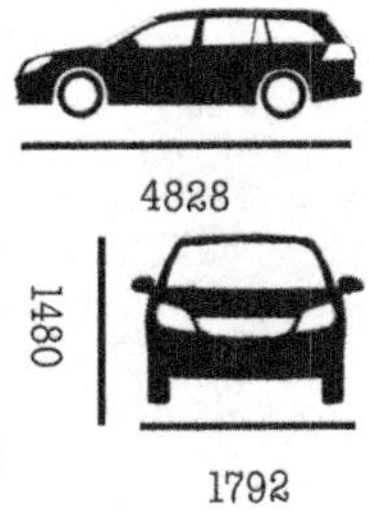

Saab 9-5 Estate 1998 - 2010

Split/fold rear seats obviously, with a flat floor which slides out beyond the bumper to help shoving stuff in. 2.0 and 2.3 and top of the range V6. 2001 revisions outside and in with diesel engines. 2005 facelift amounted to headlamps with goth level mascara, still stylish and practical. Coil pack failure, air con issues, automatic 'box + suspension bushes.

Engines: 2.0 • 2.3 • 1.9D • 2.0B • 2.3B **MPG**: 31.7-41.5. **Gearbox:** Man/Auto **Boot:** 416/1490 **Seats**: 5 **Doors:** 5

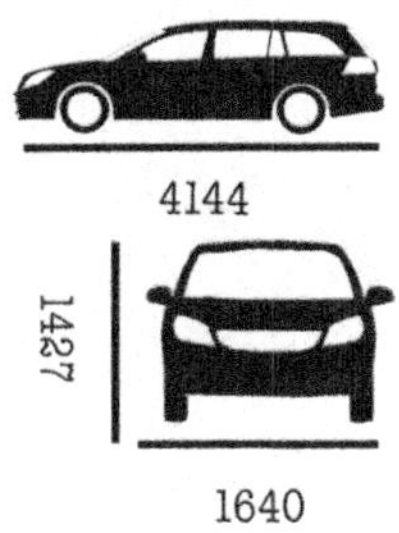

SEAT Cordoba Vario 1998 - 2001

Polo related pocket sized hatch with a big square load bay and ready to be overstuffed. 1.9 diesel is what an oil burner should be, economical and reliable, otherwise the 1.6 petrol is fine as it doesn't weigh too much. Revamp in 1999 so models looked more alike. Oil leaks, suspension wear, mass air flow sensor failure. Rust making itself known.

Engines: 1.6 • 1.9D **MPG**: 36.2-57.5 **Gearbox:** Man **Towing:** 1000 **Boot:** 390/1250 **Seats**: 5 **Doors:** 5

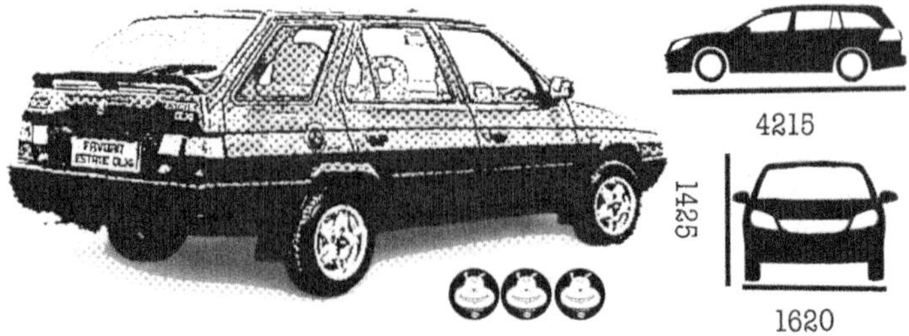

Skoda Favorit Estate 1988 - 1995

Czech new era that got better once the Volkswagen Group involved. The Germans upped the quality, revised the suspension, interior and tweaked engines. Properly basic but from '94 alloys, removable rear seat and rear wash/wipe, but there is Forman Van if required. Oily engines that get smoky, sticky gearboxes and poor electrical earths.
Engines: 1.3 **MPG**: 39.3-41.3 **Gearbox:** Man **Weight:** 920 **Boot:** 400/1300 **Seats**: 5 **Doors:** 5

Skoda Felicia Estate 1995 - 2001

VW's Skoda spin that cured overheating issues and boosted the quality quotient to make the best budget but of the '90s. Spacious and now with decent 1.6 and 1.9D VW engines. Restyled a bit in '98 and GLXi has glass sunroof, central locking, velour upholstery and a rev counter. 1.3 engine a bit creaky, worn steering, brakes and busted central locking.
Engines: 1.3 • 1.6 • 1.9D **MPG**: 38.2-49.8 **Gearbox:** Man **Weight**: 1060 **Boot:** 477/1366 **Seats**: 5 **Doors:** 5

Skoda Octavia Estate 1998 - 2005

It has metal guide rails in the floor so you can slide stuff in and it is a metre wide. There was a very broad range with high performance vRS stuff, best to stick with the basic 1.9 diesels and 1.6 to 2.0 petrols which are easy to deal with. Diesel ECU failure, water pumps, air mass sensors and ignition coils. Electrical failures in the cockpit.

Engines: 1.4 • 1.6 • 2.0 • 1.9D • 2.0D **CO2:** 143-168g/km **MPG**: 40.4-53.3 **Gearbox:** Man/Auto **Boot:** 560/1620 **Seats**: 5

Subaru Impreza Sportswagon 1993 - 2000

Apart from the turbo nutter rally one there is the Billy Basic dumpy little wagon, so you can get a 1.6 as a RWD as well as 4WD. LX has power steering, GL has electric windows, central locking. A good idea for towing, better mpg than a weighty SUV. Incredibly reliable and the only downside has been the cost of getting spares.

Engines: 1.6 • 1.8 • 2.0 **MPG**: 31.5-34.8 **Gearbox:** Man/Auto **Boot:** 356/1256 **Seats**: 5 **Doors:** 5

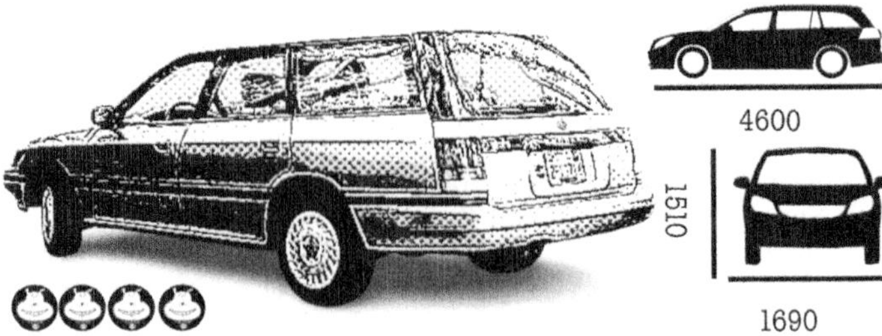

Subaru Legacy Estate 1988 - 1994

Four wheel drive estate format made the most sense and this was the farmer's posh runabout. Incredible performance on road and slightly off. Basic spec Ecos are frill free, 4-Cam is the mad one. Not very exec really. Everything costs a fortune if broken. Wear and tear on the running gear and neglect if owned by someone outside rural demographic..

Engines: 1.8•2.0•2.2 **MPG**: 27.8-29.3 **Gearbox:** Man/Auto **Weight:**1345 **Boot:** 420/700 **Seats**: 5 **Doors:** 5

Subaru Legacy Estate 1993 - 2003

Suspension re-engineered for a flatter floor, so a nice load area that gets bigger with the rear seats folded. Fairly big and thirsty petrol engines, 2.0, 2.5 and 3.0. There were some really old American built 2.0 models coming to the UK in 1999. A 2001 restyle didn't make it any prettier. Everything costs a fortune if broken. Wear and tear on the running gear.

Engines: 2.0•2.5 **CO2:** 215-228g/km **MPG**: 30.4-31.4 **Gearbox:** Man/Auto **Boot:** 433/1575 **Seats**: 5 **Doors:** 5

Suzuki Baleno Estate 1999- 2005

The only distinguishing feature is the low loading height and tailgate as part of the rear bumper. What it has going for it is Suzuki attention to detail and reliability. Mostly 1.6 petrol, but later a 1.8 engine also in the mix and not really much to recommend the Baloney except that it is a competent small family car. Rusty and the usual wear and tear issues.

Engines: 1.6 • 1.8 **CO2:** 171-184g/km **MPG**: 36.7-39.8
Gearbox: Man/Auto **Boot:** 374/1376 **Seats**: 5 **Doors:** 5

Toyota Corolla Estate 1987 - 1992

6th generation brought lots of bodywork options the station wagon was a decent holdall, if quite a rare sight. Actually a good vehicle to own and drive and not the deadly dull old thing history might have you believe. Power steering standard in 1990 and fairly classy, unlike later incarnations. Driven into the ground, yet still survive.

Engines: 1.3 • 1.6 **MPG**: 32-40 **Gearbox:** Man **Boot:** 463/508
Seats: 5 **Doors:** 5

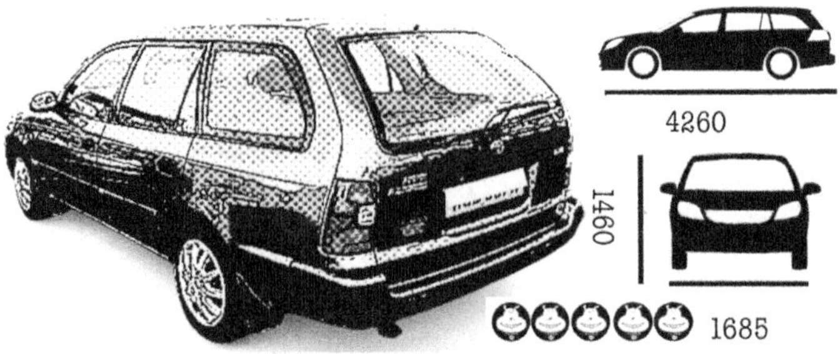

Toyota Corolla Estate 1992 - 1997

Slightly less interesting than the last one, but continues to be the small family estate of choice. Entry 1.3 engine struggles making the 1.6 engine the best all rounder. An awful lot of specs to choose from the GXi has ABS, alloys and fog lamps. Revised in '95 and there's a ton of special editions. Obviously these last forever. Most dead by rust.

Engines: 1.3 • 1.6 • 1.8 • 2.0D **MPG**: 34.6-41.9 **Gearbox:** Man/ Auto **Boot:** 394/713 **Seats**: 5 **Doors:** 5

Toyota Corolla Estate 1997 - 2002

Nothing to really set what is a compact model apart from rivals. Being a Corolla though means that it will be exceptionally reliable. Powered by unbreakable small petrols and a diesel. Avensislike styling from 2000 and a fancy common rail diesel. Rather rusty some of them, lambda sensors possibly. Mostly just worn out parts issues.

Engines: 1.4 • 1.6 • 2.0D **CO2:** 156-168g/km **MPG**: 40.4-48.7 **Gearbox:** Man **Boot:** 308/663 **Seats**: 5 **Doors:** 5

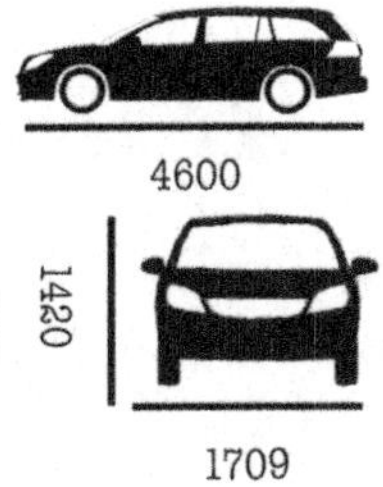

Toyota Avensis Tourer 1998 - 2003

Brit built Avensis is fabulously reliable, if dull and here was an estate, but with an awful lot more safety equipment. These are proper company cars that led hard lives but still managed to not break. There is a decent amount of room, especially with the rear seats folded. Clutch and gearbox failures over time, brake fluid leaks, plus wear and tear.

Engines: 1.6 • 1.8 • 2.0 • 2.0D **CO2:** 158-181g/km **MPG**: 36.7-47.9 **Gearbox:** Man/Auto **Boot:** 530/1000 **Seats**: 5 **Doors:** 5

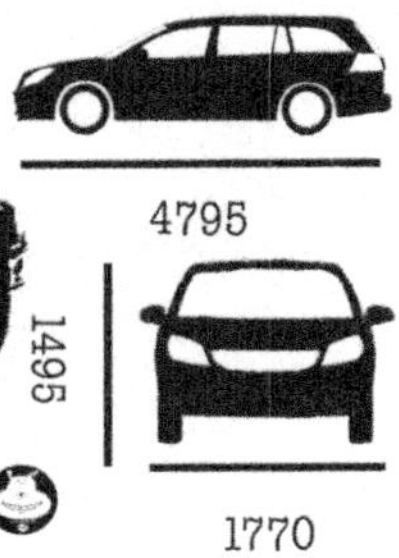

Toyota Camry Estate 1992 - 1996

Big station wagon in the States, bought because it's packed with kit and delivers utter reliability. Bags of room, and the 3.0 V6 GX has leather, cruise and headlamp washers. ABS and airbag from '93 so a pretty safe package. Very rare, the ones you find have wear and tear, otherwise it is brakes, tyres and suspension parts. Always expensive to fix.

Engines: 2.2 • 3.0 **MPG**: 24.8-32.4 **Gearbox:** Auto/Man **Boot:** 553/834 **Weight**: 1575 **Seats**: 5 **Doors:** 5

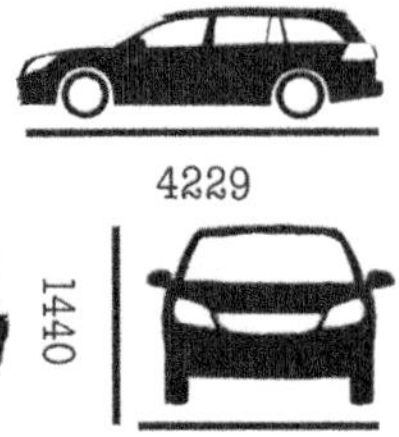

Vauxhall Astra Estate 1984 - 1991

Great range of engines and the slogging diesel delivered amazing economy for the low level rep who needed an estate. A popular fleet and family vehicle which was cheap to fix and easy to look after. Worn, brakes, suspension and smoking diesels. Most have been driven to destruction and none were cherished, pretty much extinct now.

Engines: 1.2 • 1.3 • 1.4 • 1.6 • 1.8 • 2.0 • 1.6D • 1.7D **MPG**: 34-48 **Gearbox:** Man **Boot:** 470/720 **Seats**: 5 **Doors:** 5

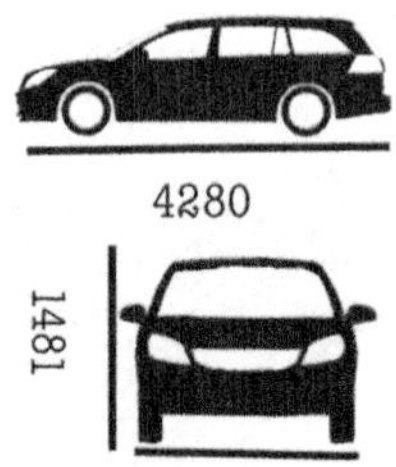

Vauxhall Astra Estate 1991 - 1998

Very much more of the same and quite right too, dated but reliable. 16 valve engine a lot more lively, but the dreary 8v 1.4 will get you there and the grim 1.7D without using much fuel at all. August '94 the V grilles with power steering, uprated suspension and other mods. CDX from '95 has alloys and CD. Worn engines, clutch and suspension.

Engines: 1.4 • 1.6 • 1.8 • 2.0 • 1.7D **MPG**: 34.1-49.1 **Gearbox:** Man/Auto **Boot:** 500/1200 **Seats**: 5 **Doors:** 5

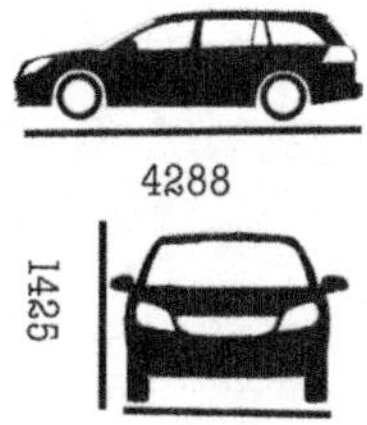

Vauxhall Astra Estate 1998 - 2004

Load bay is low wide and flat with plastic floor rails to make sliding baggage on board easier and a ski hatch to keep the rear seats upright. Massive range, seven engines. Easy to drive and live with, but open to company car abuse and can be worn out if used as a van. Corroded ECUs, cockpit leaks, failing Air conditioning and water pumps.

Engines: 1.4 • 1.6 • 1.8 • 2.0 • 1.7D • 2.0D **MPG**: 38.2-49.6 **Gearbox:** Man/Auto **Boot:** 370/1210 **Seats**: 5 **Doors:** 5

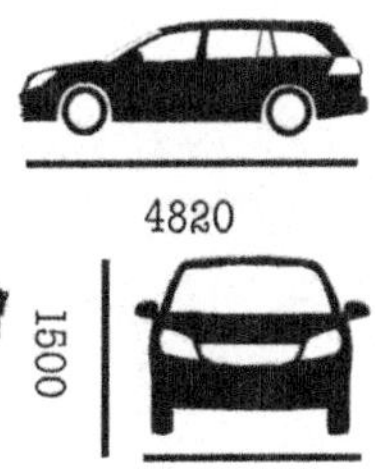

Vauxhall Omega Estate 1994 - 2003

Made in Germany with a truly massive boot, so they were well screwed together and had a Griffin badge rather than an Opel one. Better looking from 1999 and the more powerful DTi diesels which are the ones to have. A motorway missile with plenty of equipment, especially top end Elite and CDX. Worn suspension bushes, oil leaks, fuel injector seals, air conditioning and heating system.

Engines: 2.2 • 2.6 • 3.2 • 2.2D **CO2:** 192-284g/km **MPG**: 23.9-39.8 **Gearbox:** Man/Auto **Boot:** 541/1801 **Seats**: 5 **Doors:** 5

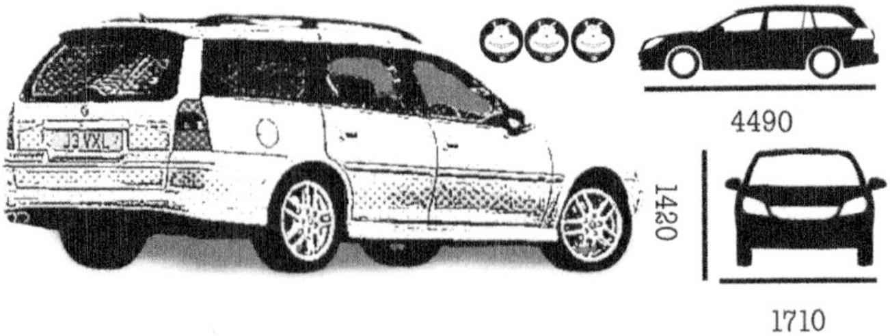

Vauxhall Vectra Estate 1996 - 2002

Big tailgate and opening with folding rear seats and all that. There are loads of engine and spec options from DTI diesels to V6 petrol. Design model has all the basics which includes cargo net, ski hatch, plus an under floor compartment. Rust obviously, leaking coolant, variously broken suspension parts, failing water pumps. Clutch a complicated job.

Engines: 1.6 • 1.8 • 2.2 • 2.6 • 2.0D • 2.2D **MPG**: 28.8-46.3 **Gearbox:** Man/Auto **Boot:** 460/1489 **Seats**: 5 **Doors:** 5

Vauxhall Carlton Estate 1986 - 1993

A big family estate car, by 1990 ABS on the petrols, the Ls and GLs don't have much, so better get a CD and CDX. Diplomat and Diamond packs deliver upgraded trim of course. Non Turbo diesel a slug and 3000 'Lotus' Lugger. Not much loved at the time, so survivors will be worn out, suspension, brakes, engine and electrics.

Engines: 1.8 • 2.0 • 2.6 • 3.0 • 2.3D **MPG**: 24.4-41.6 **Gearbox:** Man/Auto **Boot:** 530/930 **Seats**: 5 **Doors:** 5

Volkswagen Golf Estate 1994 - 1999

Blobby estate styling and not that well put together, the only lemon shaped Golf. Despite the moans another massive hit. Constantly updated and the bigger petrols engines always better, everything is average but has ultimately proved quite durable. Unloved so smoky diesel, clattery petrols and general wornoutness. Quite rare these days
Engines: 1.8 • 2.0 • 1.9D **MPG**: 36.6-56.8 **Gearbox:** Man/Auto **Boot:** 467/1500 **Seats**:5

Volkswagen Golf Estate 1999 - 2004

Nice wide and flat load area. 60:40 rear seat split, remove the headrests for a flat floor. Has classy chrome restraining hooks and stainless steel protecting the bumper and deep tailgate. 12V power output. Standard roof rails. Practical quality package. Rust on sills and arches. Engine ECU cause rough running. Brake calipers and water pump failure.
Engines: 1.4 • 1.6 • 1.8 • 2.0 • 2.3 • 1.9D **MPG**: 30.4-57.6 **Gearbox:** Man/Auto **Boot:** 460/1470 **Seats**: 5 **Doors:** 5

Volkswagen Passat Estate 1988 - 1996

The last truly dull and boring one, so a great estate. Incredible reliability and ability to rack up massive business mileages. Superb engines better to have a 2.0 petrol or the 19TDi. Incredibly there's a VR6. Late life revamp in 1994 so there's a restyle and spec upgrades with ABS on the big engines. High mile wear & tear to engine/suspension.

Engines: 1.8 • 2.0 • 2.8 • 1.6D • 1.9D **MPG**: 28-53.3 **Gearbox:** Man/Auto **Boot:** 465/1500 **Seats**: 5 **Doors:** 4

Volkswagen Passat Estate 1997 - 2001

Generously equipped, had fine engines and subtle styling with a large load area, that can obviously be extended with the split/folding seats. Easy to load, too with a conveniently low and flat lip. From the S climate control, electric sunroof, engine immobilser and ABS brakes. Smoking diesels, damaged suspension, worn brakes and steering

Engines: 1.8 • 2.0 • 2.3 • 2.8 • 1.9D • 2.5D **MPG**: 26.9-51.4 **Gearbox:** Man/Auto **Boot:** 475/1500 **Seats**: 5 **Doors:** 5

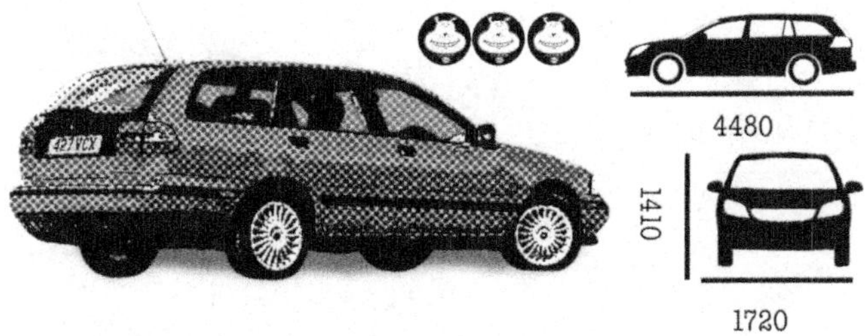

Volvo V40 1996 - 2004

A pokey estate for a Volvo. Still successful though, most buyers went for the 1.9D which is an excellent Renault Turbo diesel. 1.6 or 2.0 petrols are better options for the modern world. Updated in 2000 with new engines and styling. Survivors are rusty. Corroded loom causes poor electrics, air con/power steering can fail, brake calipers break. **Engines:** 1.6•1.9D **MPG**: 35.8-52.3 **Gearbox:** Man/Auto **Boot:** 368/1422 **Seats**: 5 **Doors:** 5

Volvo 240 Estate 1990 - 1993

The saloon went in 1991, leaving this old Ledge in DL, GL and GLT which had 133bhp. 240GL was the run out estate, until the 1993 240 Torslanda with some extra spec. Ones that are not antique dealer time warps will have rotten bodies and engines that are smokey, cammy and all the ancillaries go pop.

Engines: 2.0•2.3 **MPG**: 27-29 **Gearbox:** Man/Auto **Boot:** 1200/2151 **Weight:** 1305 **Seats**: 5 **Doors:** 5

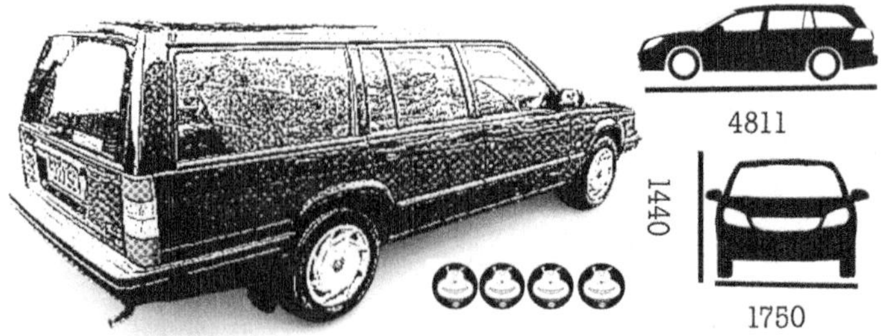

Volvo 940/960 Estate 1990 - 1998

Here is the 740/760 reinvented for the '90s, a big collection of useful rectangles. Better to drive when the suspension was tweaked in '92, then again in '94 with more chassis work and softening the styling rectangles.. Mind boggling spec/engine combos. Driven into the tarmac so worn suspension, smoky turbos and all that. Usefully gigantic.

Engines: 2.0 • 2.2 • 2.3 • 2.5 • 3.0 • 2.4D **MPG**: 24.8-36.7
Gearbox: Man/Auto **Boot:** 991 **Seats**: 5 **Doors:** 5

Volvo 850 Estate 1992 - 1996

Front wheel drive estate that was better than ever on the road and at lugging stuff.. A bunch of great 5 cylinder engines too, from adequate to outrageous and frugal. Loads of spec level combinations which were always being revise. Tired 10V 2.0. smokey diesels and worn suspension. Faithful old thing which proved itself with the Cops.

Engines: 2.0 • 2.3 • 2.5 • 2.5D **MPG**: 25.2-46.7 **Gearbox:** Man/ Auto **Weight:** 1473 **Boot:** 792/1580 **Seats**: 5 **Doors:** 5

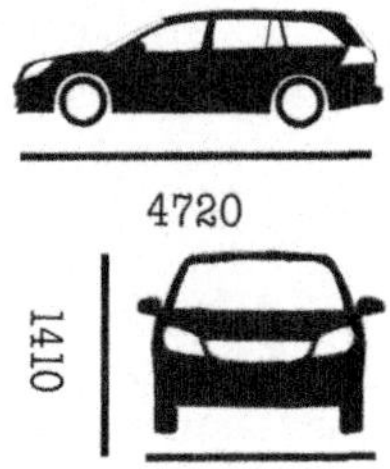

Volvo V70 1996 - 1999

The 850 was good and this is the Mark 2 version as Volvos got cuddlier. Lots of engine options and madder R and T5 plus loads of option packs including winter, Comfort, SE and XT. Confusing, but at least whatever combo you end up with it is at least solidly reliable. The usual high mile issues that means rattles and smoke.

Engines: 2.0 • 2.4 • 2.5 • 2.5D **MPG**: 24.1-44.8 **Gearbox:** Man/ Auto **Boot:** 480-610 **Weight**: 1459 **Seats**: 5 **Doors:** 5

Space Ships - Banger Vans

This is the historical point at which just about every motor manufacturer believed that they needed to offer a People Van in their line up. This is when they were really born. Never mind the Renault Espace from the 1980s and those original American efforts in the shape of the Dodge Caravan and its related Plymouth Voyager, those were the school run and soccer mom, vehicles of choice Stateside.

Europe were slow to see the point. If someone needed to use a minibus they hired one and it would be a Ford Transit, Morris or Leyland Daf Sherpa. Excitingly it might even be an ancient Commer Spacevan. Certainly you could get lost down the actual van with windows and some seats rabbit hole and perhaps I will one day.

The Renault Espace was an oddity, but a stylish one which was adopted by the well off with more than three sprogs. For the more cash strapped in the '80s there was the Daihatsu Hijet which was a sort of reinterpretation of the original Fiat Multipla, but probably more dangerous. Perhaps we should double back and shoehorn the equivalent Bedford Rascal into the Bangerpedia. (We did).

Meanwhile, outside of grey import Hummers, Americana did arrive and was built in Austria as the Chrysler Voyager. The Honda Shuttle was something they liked Stateside, the Mazda MPV was appropriately badged and even those narrow gauge Nissan Serenas could squeeze eight on board. The Toyota Previa though really was the size of a space shuttle and capable of just as many miles.

Mostly what we got in the '90s was large family vehicles that were like cars to drive rather than big, boomy and coarse vans with seats screwed to their floors. Actually there were some decent adapted vans, see further the Mercedes range of V Wagens.

Unless you need a Space Ship for your large family or are considering a spot of private hire work, these are bloody awful. things.

Bedford Rascal 1986 - 1993

Later badged as a Vauxhall (1990) joint venture between General Motors and Suzuki (Super Carry) and built in Dunstable. Most were vans, some became campers, a few had seats screwed into the payload area and became the cutest people carriers on the planet. Rust has killed most. Smoke and oil leaks and general neglect finished them..

Engines: 1.3 **MPG**: 38.0 **Gearbox:** Man **Payload:** 550kg **Weight:** 755 **Seats**: 6 **Doors:** 5

Chrysler Voyager 1997 - 2001

This does seem to be roughly the size of a planet without actually being a van with windows. There was an even grander, Grand model. Everything as standard specification. Not just large American related petrols, also an Italian made VM Turbo diesel. Oil leaks, automatic gearbox, diesel turbo smoke and rust.

Engines: 2.0 • 2.4 • 3.3 • 2.5D **CO2:** 185-297g/km **MPG**: 22.2-36.7 **Gearbox:** Auto **Boot:** 445/4080 **Seats**: 5/7**Doors:** 5

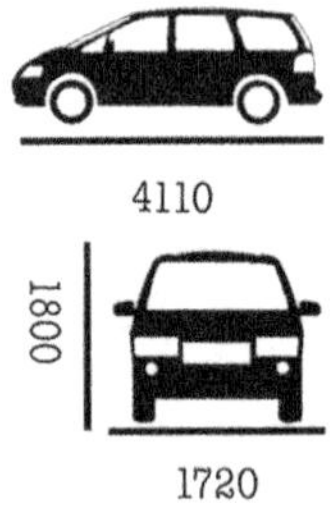

Citroen Berlingo 1998 - 2009

Endearing little van. Flexispace models have extra cubbies and also roof shelf areas. Not a lot of standard items on early models. Usually cheap to run whether as a diesel or small petrol. Actually fun to drive, beware ones actually used as vans. Suspension repairs expensive, cabin and under bonnet water leaks, ECU malfunction. Complicated wiring. Rust.

Engines: 1.4 • 1.6D **CO2:** 150-195g/km **MPG**: 34.5-49.6
Gearbox: Man/Auto **Boot:** 624/2800 **Seats**: 5 **Doors:** 5

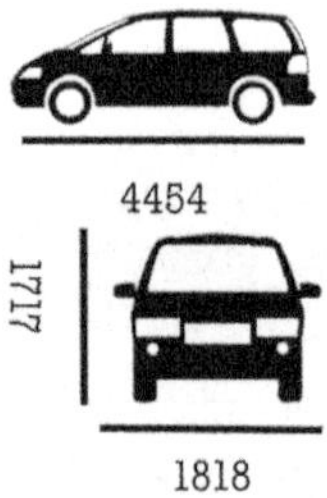

Citroen Synergie 1995 - 2002

One of three, see also Peugeot and Fiat. A box with seats which went from six to seven and eventually eight depending on the model. All upgraded for '98 and overall the specification was fine even though this was 'cheapie' of all the van versions. Leaking coolant, clutch failure, ECU issues and MOT failure level rust.

Engines: 2.0 • 2.0D **CO2:** 182-232g/km **MPG**: 30.7-42.2
Gearbox: Man/Auto **Boot:** 340/3313 **Seats**: 5/7 **Doors:** 5

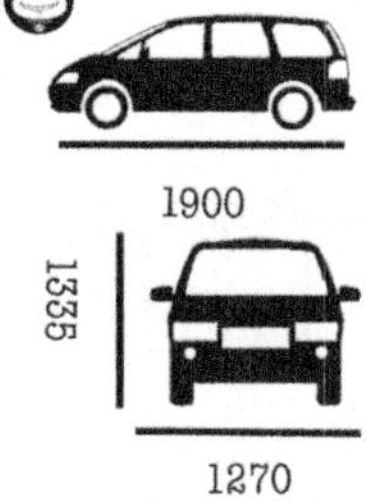

Daihatsu Hijet 1986 - 2004

A real micro van with more seats than are truly comfortable or safe. Considering its size the 1.3 engine is more than enough. Will seat six incredibly. Better enjoyed as a van, tipper or pick-up. Seem to buzz around well beyond their recorded time of death. Enjoy. Rust have killed most. Smoke and oil leaks and general decrepitude.

Engines: 1.3 **CO2:** 159g/km **MPG**: 38.0 **Gearbox:** Man **Boot:** 560kg+ **Seats**: 6 **Doors:** 5

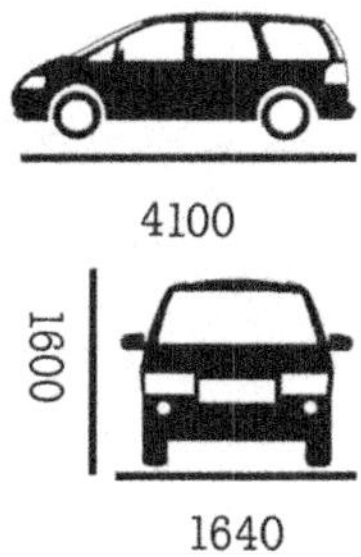

Daihatsu Gran Move 1997 - 2001

Is this a people mover? Maybe not, it is a tall mini estate. It sold on value and the fact that it seated 5 reasonably spaciously with some luggage. Surprisingly safe with airbags, pretensioners and ABS on a Move+ which also has alloys, heated mirrors and load cover. 1600 engine from late '98. Engine very reliable some parts hard to find.

Engines: 1.5 • 1.6 **MPG**: 37.2 **Gearbox:** Man **Boot:** 400/1490 **Turning**: 9 **Weight:** 965 **Seats**: 5 **Doors:** 5

Fiat Ulysse 1995 - 2003

There's a Citroen and Pug version, these were the ones which always ended up as the cheapest in the classifieds. Still the case for any survivors. Seat options are from five to eight and most of them perform lots of swiveling, sliding turning tricks. Remember it is a Fiat. Structural rust, oil leaks, ECU failure, oil leaks and smokey exhaust.

Engines: 2.0 • 2.0D **CO2:** 182-221g/km **MPG**: 30.7-42.2 **Gearbox:** Man/Auto **Boot:** 340/3313 **Seats**: 5/7 **Doors:** 5

Ford Galaxy 1995 - 2000

The nation's absolutely favourite people shifter. Not the cleverest of all MPVs, just decent to drive, practical and spacious enough. Good engines from the VW diesel to the creamy V6 and Ford 2.3 petrol compromise. Reliability got better. Oil leaks, turbocharger wear. MOT failure structural rust. Corroded brake pipes, air con and gearbox failures.

Engines: 2.0 • 2.3 • 2.8 • 1.9D **MPG**: 28.0-42.8 **Gearbox:** Man/Auto **Boot:** 810 **Seats**: 5/7 **Doors:** 5

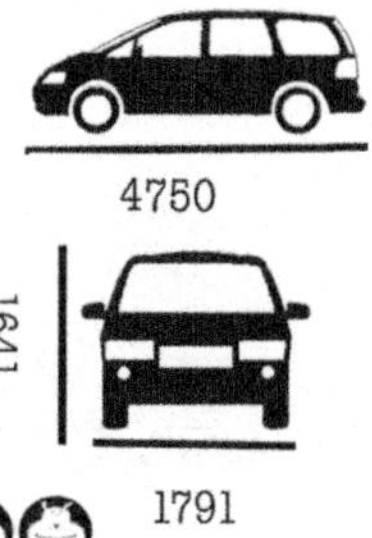

Honda Shuttle 1995 2001

Accord based, so comfy and quality finish. Middle seats difficult to hoof out and rear seats are hard benches. Decent standard kit SE has alloys, cruise, roof rails, new line up with 2.3 from '98. and a reputation for not having any mechanical sulks. Worth finding. Rust on the old ones, parts are difficult to find, otherwise just wear and tear.

Engines: 2.2 • 2.3 **MPG**: 26.6-27.6 **Gearbox:** Auto **Boot:** 278/2902 **Weight**:1465 **Seats**: 6/7 **Doors:** 5

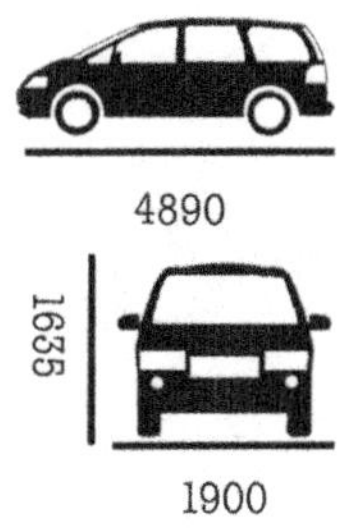

Kia Sedona 1999 - 2006

Effectively the people's Carrier. An awful lot of seats and room for the money. Lots of kit. Captain's swivel chairs in the second row don't just swivel, they slide as well. Bed recliners and double figure cupholders, including rear air con controls. MOT failure level structural rust to floor. Electrics ABD sensors and ECUs.

Engines: 2.5 • 2.9D **CO2:** 217-261g/km **MPG**: 25.7-34.4 **Gearbox:** Man/Auto **Boot:** 588/3616 **Seats**: 7 **Doors:** 5

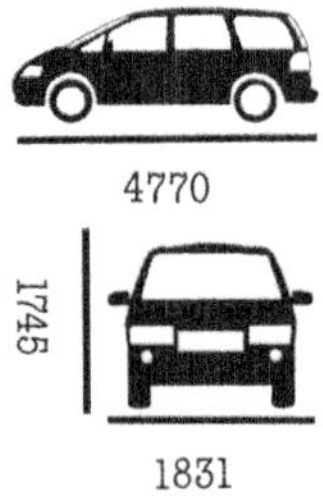

Mazda MPV 1999 - 2004

It is just a seven seat bus with sliding doors on both sides of the bodywork, which also contains Mazdareliability, which is more than enough when it is banging about later in life. Lots of standard kit and the usual seatage stuff, sliding folding and removing. Wear and tear now on these, rust on the earliest ones, brake discs and suspension bushes.

Engines: 2.0•2.3•2.0D **CO2:** 188-237g/km **MPG**: 28-39.8
Gearbox: Man **Boot:** 294/866 **Seats**: 5 **Doors:** 5

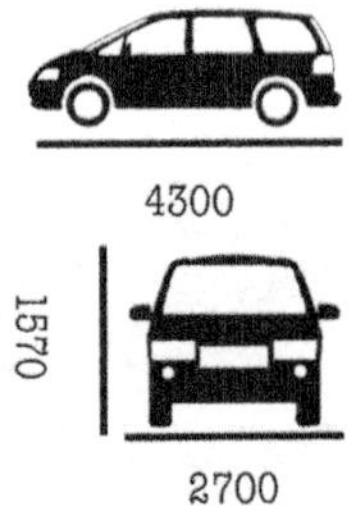

Mazda Premacy 1999 - 2004

Takes Five. Those seats do all the usual, recline, fold flat, hinge forward. Good to drive, if that really matters. All that matters is Mazdareliability. The standard specification is pretty good, if you want any more than the GSi delivers. This is a bit like a Ford Galaxy, but less useful. Won't break down though, light rust, wear and tear.

Engines: 1.8•2.0•2.0D **CO2:** 177-209g/km **MPG**: 33.6-49.6
Gearbox: Man/Auto **Boot:** 371/719 **Seats**: 5 **Doors:** 5

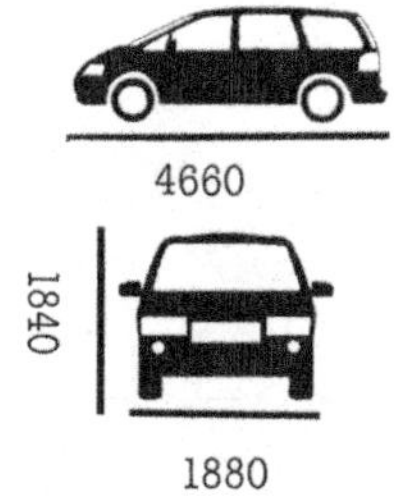

Mercedes V-Class 1996 - 2002

Vito van, room for 7 and whopping sliding doors with S luxury Class touches. Here is a utility vehicle, for the posh people. Diesel or petrol, depends how well off the buyer is. Ones that are used by the hire industry might be tired, also pretty rusty by now. Suspension and brakes have to be checked also tired diesels, leaking fluids etc.

Engines: 2.3 • 2.8 • 2.2D **CO2:** 199-330g/km **MPG**: 20.5-35.5
Gearbox: Auto **Boot:** 580/4559 **Seats**: 6/7 **Doors:** 5

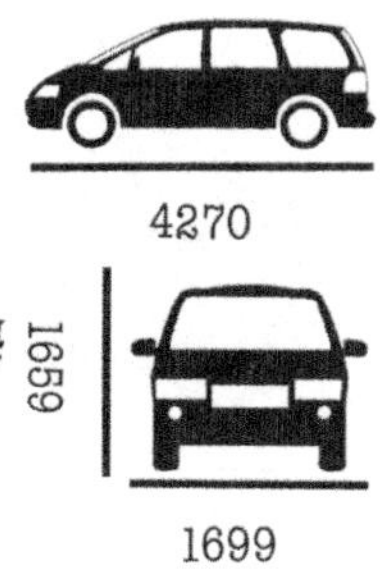

Mitsubishi Space Wagon/Runner 1991 - 1999

Runner seats 5 and the Wagon manages 7. The trouble is that it isn't that flexible and with 7 up, not any meaningful luggage space. That makes these just a well built estate car with a high roof. Split/fold rear seat, Diamond Pack adds glass sunroof, roof rail and luggage cover. Suffers from rust, wear and tear, check exhaust and brakes.

Engines: 1.8 • 2.0 • 2.0D **MPG**: 30-6 -32.8 **Weight**:1228
Gearbox: Man/Auto **Boot:** 521 **Seats**: 5/7 **Doors:** 5

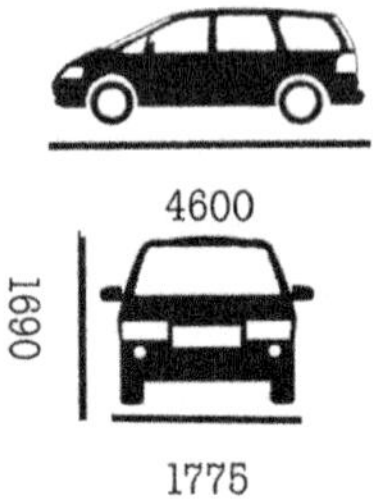

Mitsubishi Space Wagon 1999 - 2004

Third incarnation is larger, but still more estate than van like. Seven seats in three rows and that second row is an old fashioned bench. Top of the range GLS has 6 separate seats allowing passengers to walk through. Folding and removing are also options. Yes its all about space. Suffers from rust, wear and tear, check exhaust and brakes..

Engines: 2.0 • 2.4 **CO2:**225-229g/km **MPG**: 29.7-28.0
Gearbox: Ma/Auto Boot: 241/1931 Seats: 7 Doors: 5

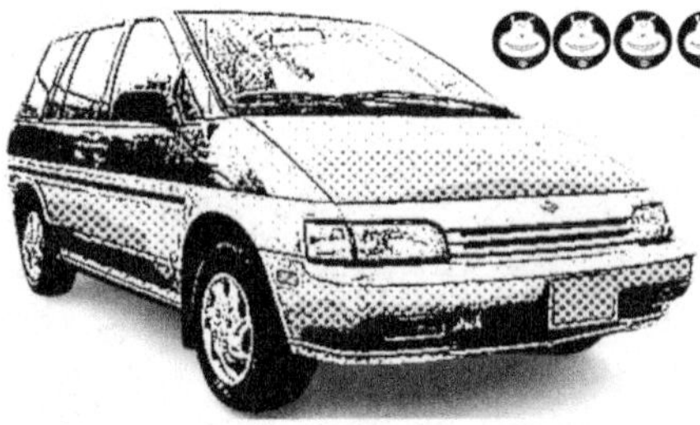

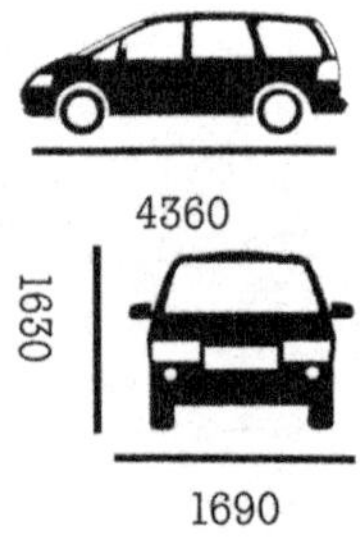

Nissan Prairie 1989 - 1991

Surprisingly short lived people van that looked the bees stylish knees, with sliding door convenience and tons of room inside. Usually front wheel drive, 4WD Japanese market models available for that go anywhere ability that UK buyers shunned. Super reliable and Bullet Train styling are the main plus points. Grey imports more common.

Engines: 2.0 • 2.4 **MPG**: 29-30 **Gearbox:** Man **Boot:** 250/1800 **Seats**: 5-7 **Doors:** 5

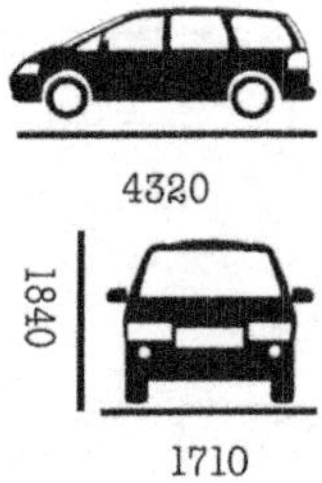

Nissan Serena 1993 - 2001

It was a narrow Japanese market van with added seats. Quite ancient tech, but incredibly there are 8 seat versions. Seats can be removed and moved about, which is no longer a novelty. Old school Nissaness means that it they are really tough. Rot kills them eventually. Some big door mirror imports around. Rust, worn interior/exterior.

Engines: 1.6 • 2.0 **CO2:** 244g/km **MPG**: 25.9-28.2 **Gearbox:** Man/Auto **Boot:** 350/1900 **Seats**: 5/8 **Doors:** 5

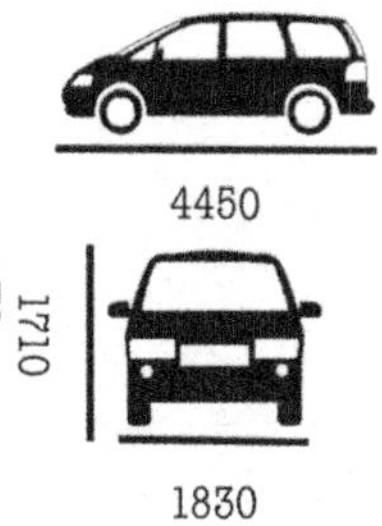

Peugeot 806 1995 - 2002

Familiar because it is also a Citroen and a Fiat. This Pug was also available with 5 or 7 seats. It is just a big old fashioned box, but still pretty useful. The seats do the usual, so that's recline, swivel fold and remove. These aren't looked after and can be quite the worse for wear. Rust, leaky sunroofs, ECUs, smoky diesels, most things really.

Engines: 2.0 • 2.0D **CO2:** 182-221g/km **MPG**: 30.7-42.2 **Gearbox:** Man/Auto **Boot:** 340/3313 **Seats**: 7 **Doors:** 5

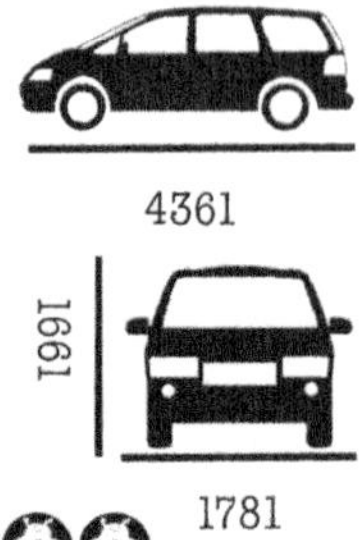

Renault Espace 1984 - 1991

The original model, a van for families, not completely original but certainly the most stylish. Apart from the driver's seat, all the others could be removed. The fronts even swiveled 180 degrees, Captain style. The second and last row of seats could have been reclined and form a king-size bed. Clutches can have a short life, smokey diesels.
Engines: 2.0 • 2.2 • 2.9 • 2.0D **MPG**: 26.4-35.8 **Gearbox:** Man **Boot:** 580/3060 **Seats**: 5/7 **Doors:** 5

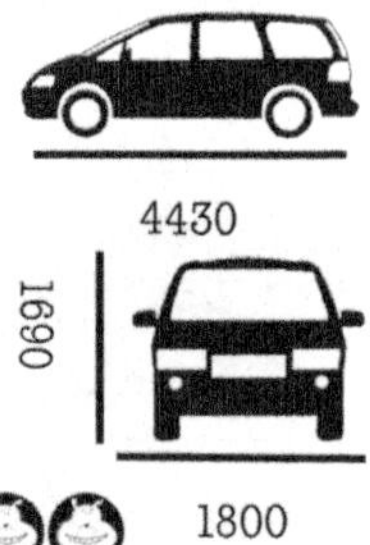

Renault Espace 1991 - 1997

Revamped premium people mover consolidates its success as an interior packaging masterclass. The seats are multi adjustable, there are stowage boxes and big door pockets. Otherwise not a sharp drive, but the diesel is good and the V6 fine if you can afford it. Driver's airbag from '95. Clutches can have a short life, smokey diesels and failing ECUs.
Engines: 2.0 • 2.2 • 2.9 • 2.1D **MPG**: 26.4-37.4 **Gearbox:** Man/ Auto **Boot:** 900/3000 **Weight:** 1496 **Seats**: 7 **Doors:** 5

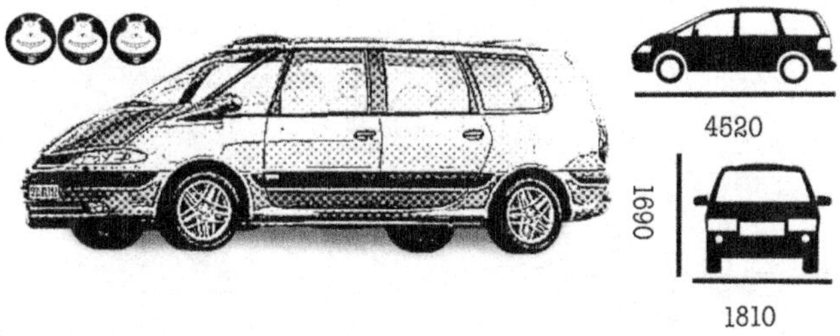

Renault Espace 1997 - 2003

MPV royalty. Sophisticated third version that is spacious and versatile, usually with decent standard kit. Got the removable and folding seats and the higher spec RXEs have rail system for the seats to slide on. It is a Renault, so the electrics can be flakey especially if an electric handbrake is fitted. Clutches can have a short life, smokey diesels and failing ECUs.

Engines: 2.0•3.0•2.2D• **CO2:** 189-273g/km **MPG**: 24.4-39.8 **Gearbox:** Man/Auto **Boot:** 600/2600 **Seats**: 7 **Doors:** 5

Renault Scenic 1997 - 2003

Not just a Megane anymore, now its own life form. Essentially an overgrown hatch with seats on rails, air con extended to glove box and posh models had spilt tailgate as well. A ton of trim levels and engines, Renault fragility and electrics can spoil it all. Sensors and various ECUs, coil packs, clutch, heater matrix, rust.

Engines: 1.6•1.8•2.0•1.9D **CO2:** 165-224g/km **MPG**: 29.7-49.6 **Gearbox:** Man/Auto **Boot:** 411/1798 **Seats**: 5 **Doors:** 5

Renault Kangoo 1999 - 2008

Vannish rural France appeal in the mould of ancient Renault 4s. Not a luxury interior, it's a hose down affair. The rear seat pivots forwards so there is instant load space. Lots of extras available to fancy it up and many buyers did just that. Electrics from basics like wipers, ignition coil packs, automatics can fail. EGR valves regular Renault soft spot.
Engines: 1.2 • 1.6 • 1.5D **CO2:** 139-190g/km **MPG**: 35.3-53.3 **Gearbox:** Man/Autos **Boot:** 600/2600 **Seats**: 5 **Doors:** 5

SEAT Alhambra 1996 - 2000

The one that wasn't a Ford or Volkswagen which came with the better VW engines. Folding and removable seats if you can manage to hoof the heavy old things out. Front ones on some models do the Captain swiveling thing. Oil leaks, turbocharger wear. MOT failure structural rust. Corroded brake pipes, air con and gearbox failures.
Engines: 2.0 • 1.8 • 1.9D **MPG**: 28.2-44.1 **Gearbox:** Man/Auto **Boot:** 267/2297 **Seats**: 5/7 **Doors:** 5

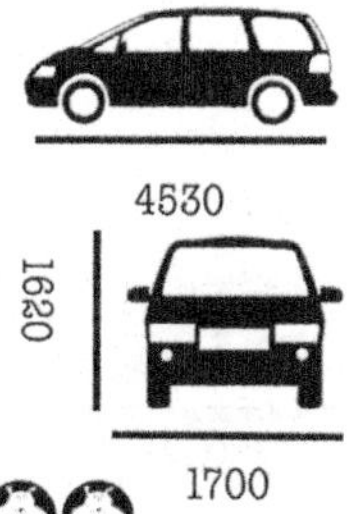

Toyota Picnic 1997 - 2001

Great name, actually called a Family Fun Vehicle by the manufacturer with funky two tone paintwork that seated six inside its compact dimensions. Optional seven seats as was dual sunroofs. Simply buy because it is a Toyota rather than a packaging masterclass. And because of the name. Old age issues, so rust, wear and tear to suspension and brake parts.
Engines: 2.0 • 2.2D **CO2:** 226g/km **MPG**: 29.1-36.1 **Gearbox:** Man **Boot:** 182/1841 **Seats**: 6 **Doors:** 5

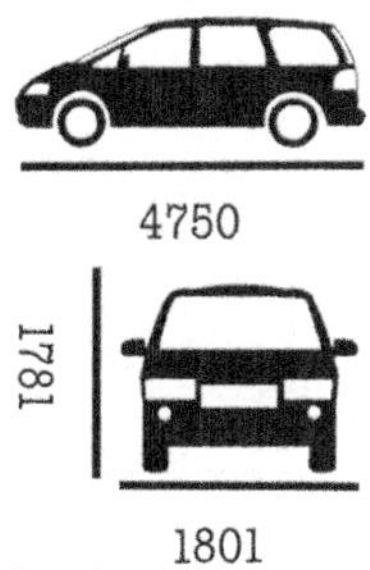

Toyota Previa 1990- 2000

Mid engined MPV no less. Reclining and sliding rear seats. Few take out the seats unless relocated to the lounge. Double bed option. Sliding door on nearside. Estima 4WD import if you need it and diesel. Mostly this is the best big family MPV of all time. Subject to wear and tear mostly, so it will be tyres, suspension bushes and brake parts.
Engines: 2.4 **MPG**: 25.2 **Gearbox:** Man/Auto **Boot:** 671 **MPH**: 106 **BHP**: 132 **Weight**: 1691 **Seats**: 7/8 **Doors:** 5

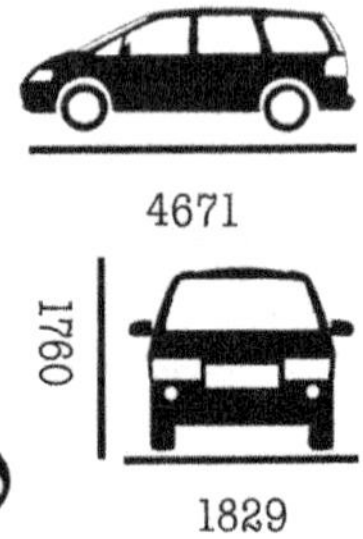

Vauxhall Sintra 1997 - 1999

Not around for long but in practical terms this Chevrolet Venture spin off whatever the seat format was ordered in had more interior space than the Galaxy related ones. Bit iffy in the NCAP crash test. Great seats and surprising amount of luggage space. V6 is probably the best, but practically the 2.2 DTi is the answer. Just going to be scruffy. **Engines:** 2.2 • 3.0 • 2.2D **MPG**: 24.6-35.5 **Gearbox:** Man/Auto **Boot:** 459 **Seats**: 5/6/7 **Doors:** 5

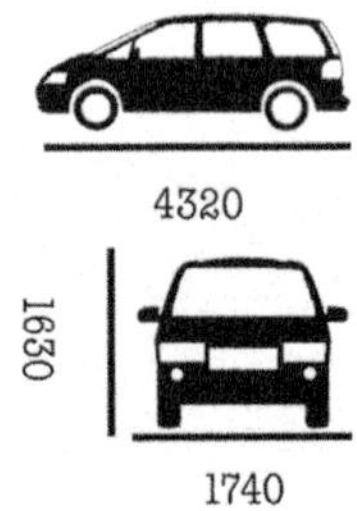

Vauxhall Zafira 1999 - 2005

Compact MPV for the masses which is an Astra with seven seats. Middle seats slide and the rearmost fold into the floor. That does create a massive amount of load space if required. A hard working and flexible people shifter that isn't a pig to drive or park. Water pump failures, seized brakes, water leaks into cabin and ECU irregularity.

Engines: 1.6 • 1.8 • 2.2 • 2.0 • 2.0D **MPG**: 27.9-43.5 **Gearbox:** Man/Auto **Boot:** 150/1699 **Seats**: 7 **Doors:** 5

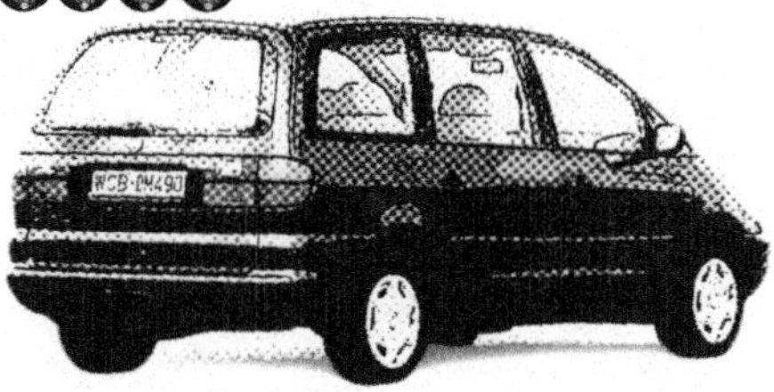

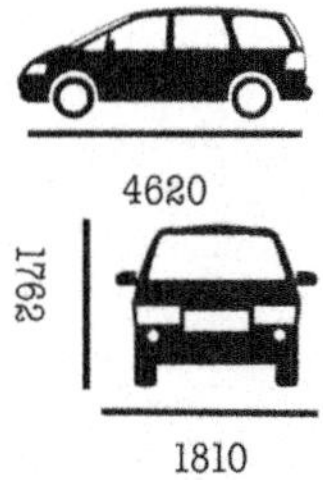

Volkswagen Sharan 1995 - 2000

Passat 2.0 engine and their usual turbodiesels are rather good. Drivers enjoy the sheer carness of it. V6 Synchro is a pointless laugh. Seats do the fold down and pull out tricks. Five or seven seats. Often worn out. Oil leaks, turbocharger wear. MOT failure structural rust. Corroded brake pipes, air con and gearbox failures.

Engines: 1.8 • 2.0 • 2.8 • 1.9D **MPG**: 25.2-44.1 **Gearbox:** Man/ Auto **Boot:** 255/2608 **Seats**: 5/7

BUVs - Banger Utility Vehicles

Here is the crossover decade. When 4 x 4s went from working vehicles to crowding up your high street and parking jauntily on the kerb. Indeed, some were calling them 'crossovers', others 'off roaders' and just a few taking on board the stateside SUV moniker.

Never mind the Landed Gentry Range Rover, here was a bunch of 4×4s for us peasants. What was interesting is the mix of old fashioned farm hands and the new breed of high rise family cars.

There were some of the Yank Tank SUVs that came straight over on the boat. Most of the Jeeps were very welcome, especially the Wrangler and the original boxy Cherokee. However, the Chevrolet Blazer was not quite as at home, or the Ford Explorer. Never mind the inspired by or built under licence Willys Jeepneys were interesting, such as the Asia Rocsta and Mahindra. That's what car buyers thought of when the word Jeep was mentioned. This also explained how the Suzuki SJ thrived for a while, but it was their Vitara which really changed everything.

Toyota knew what they were doing, hence the Land Cruiser, but the first RAV-4 was a really fun twist on the whole Jeep thing. I suppose it is when everyone started to ape the Range Rover that things started to come slightly unstuck. Except that the Mitsubishi Shogun was sensationally capable and cheaper. Land Rover realised that they had a powerful brand however unreliable their products are. In fact, the original Discovery was straightforward enough to make sense provided you did't bag a 200 Mpi.

It was a very eclectic bunch of vehicles for all budgets and tatses. So as well as the clunky Nissan Patrols was the lifestyle Honda HR-V and contributions from South Korea, be they Daewoo or SsangYong Korandos. Meanwhile Rumania was making some foreign currency with their half baked Duster. A name that would return in the next century, but not in another Bangerpedia.

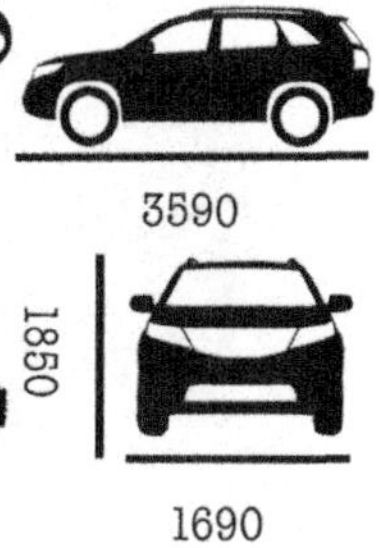

Asia Rocsta 1986 - 1997

A Jeep on the cheap, very basic, but owners swore by them. Powered by Kia engines, this is effectively the Korean army's 4×4. Available with hard and soft top and some say it was only available in the '90s, but actually dates from 1986. Not comfortable, just rather charming. An off roader as it is supposed to be. Rot and finding the right parts.

Engines: 1.8 • 2.2D **MPG**: 23.5-35.3 **Gearbox:** Man **Weight**: 1270 **Boot:** 110/470 **Towing:** 1000 **Seats**: 5 **Doors:** 3

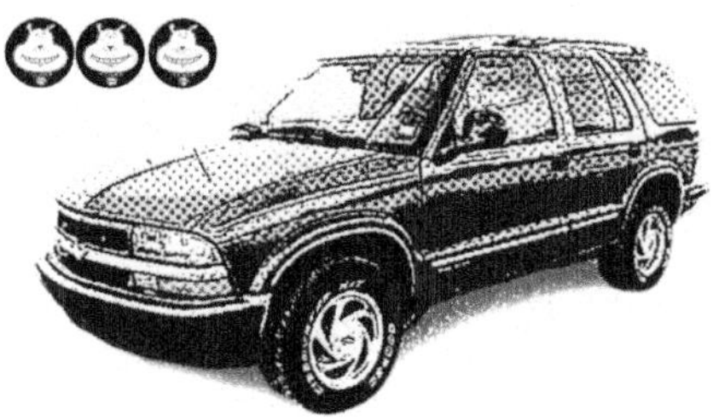

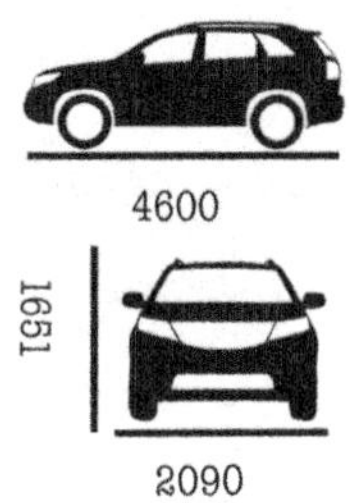

Chevrolet Blazer 1999 - 2002

Very old school 4×4 with quite a soft suspension set up which comes with all the basic luxury spec items which seemed lush in the '90s. Just the five seats and V6 petrol, but you'd buy it as a hack or for towing. Has the proper high and low range gearbox settings. By this stage in its life could be quite corroded. Engine should be fine.

Engines: 4.3 **CO2:** 345g/km **MPG**: 19.1-20.6 **Gearbox:** Man/ Auto **Boot:** 1056 **Towing:** 2041 **Seats**: 5 **Doors:** 5

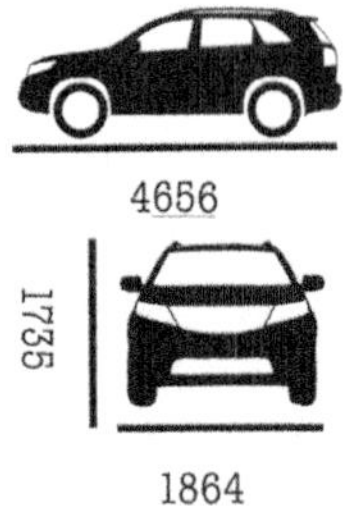

Daewoo Musso 1999 - 2002

Previously home to the tongue twisting SsangYong badge. A quirky Korean which is tough as an everyday tow car with a big boot and more equipment than before. Mercedes related petrol engine and a turbo diesel are both pretty unburstable. Strictly five seats which is fine, just look out for rot and neglect. Fluid leaks and minor electricals.

Engines: 2.3 • 2.9D **CO2:** 245-282g/km **MPG**: 23.3-30.7 **Gearbox:** Man/Auto **Boot:** 1119-1920 **Towing:** 2800 **Seats**: 5 **Doors:** 5

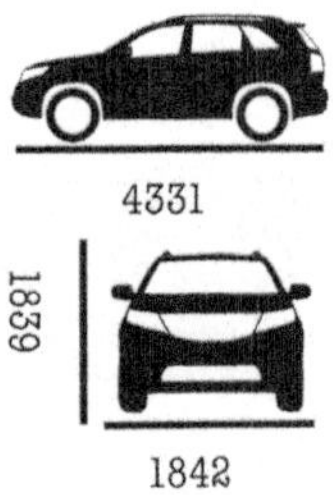

Daewoo Korando 1992 - 2002

Essentially an ugly Jeep which is yet another SsangYong leftover, this time with a Mercedes five cylinder diesel which was good. Just the three doors, but is decently big inside with a good boot. Basic 4×4 stuff which isn't bought to pose, but tow and work hard which is not great when buying used, so rust and worn out mechanicals are the only concerns.

Engines: 2.3 • 2.9D **MPG**: 27.7-28.8 **Gearbox:** Man/Auto **Boot:** 479/1464 **Towing:** 2800 **Seats**: 5 **Doors:** 5

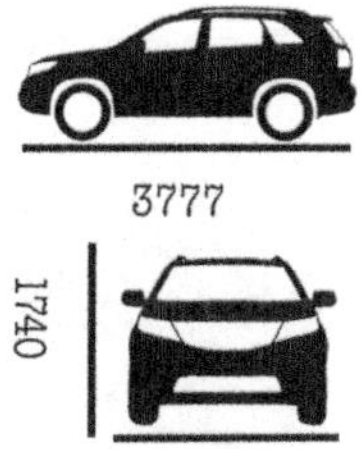

Dacia Duster 1986 - 1996

ARO 10 as it was known in Romania came to the UK through the '80s and '90s and sold on price. Renault engines and the warning that it would be unwise to venture too far off road. Enough people bought them, either the convertible or the hardtop estate for imports to continue for some time. Tatty from the off, now retro cool if any survive.

Engines: 1.4 • 1.5D **MPG**: 25.7-34.8 **Gearbox:** Man **Boot:** Yes... **Weight**: 1280 **Towing:** 1000 **Seats**: 5 **Doors:** 3

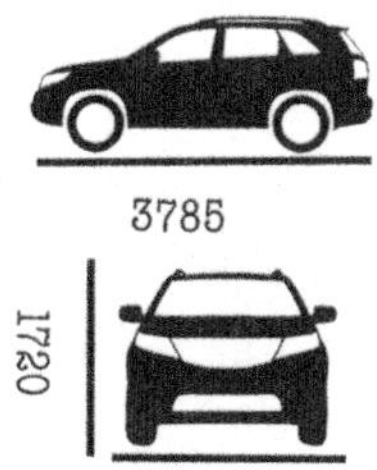

Daihatsu Sportrak 1984 - 2002

Lightweight version of the Fourtrak, sort of. Best suited to holiday hire duties somewhere sunny for the DX soft top one. A more grim location for the EL hardtop, which can certainly hack it off road. ELXi has freewheel front hubs. Restyled in '94 with revised suspension and specifications. Look for rust, wear and tear.

Engines: 1.6 **MPG**:27.8 **Gearbox:** Man **Boot:** 295/920 **Towing:** 1500 **Weight:**1206 **Seats**: 5 **Doors:** 3

Daihatsu Fourtrak 1989 - 1997

First genuine altDefender 90. Farmers and builders bought these and quite right too. Really basic until the '90s with uprated brakes and running gear. Got independent suspension in 1993 and incredibly a seven seat version. Diesels replaced the petrol. Rust, wear and tear look out for old farm hacks which could be barely alive. Great workers.

Engines: 2.8 **CO2:** 248g/km **MPG**:30.1 **Gearbox:** Man **Boot:** n/a **Towing:** 3500 **Seats**: 5 **Doors:** 5

Daihatsu Terios 1997 - 2005

Proper off roader in teeny package. Big facelift in 2000 changed the look somewhat and got the 1.3 petrol from a Toyota Yaris, so that will always last. They fiddled with the specifications by realignment of model and specs in 2003, but that's irrelevant. Yes it's really a Starlet/Yaris 4×4. Wear and tear issues only, tyres brakes.

Engines: 1.3 **CO2:** 205g/km **MPG**: 36.7 **Gearbox:** Man/Auto **Boot:** 221/561 **Towing:** 1400 **Seats**: 5 **Doors:** 5

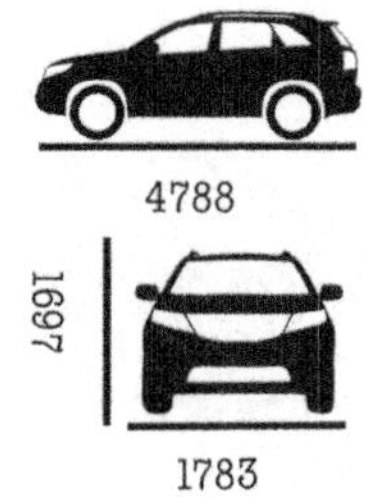

Ford Explorer 1997 - 2001

Had a reputation for rolling over, but sorted out and packaged with most electrical things as standard. Powered by a 4.0 V6 petrol it provides a comfortable ride and was pretty decent at towing. Easy to live with and getting parts is not impossible on the basis that it is a best seller. Blue engine smoke, worn suspension, broken electrics. Some rust.
Engines: 4.0 **CO2:** 350g/km **MPG**: 19.3 **Gearbox:** Auto **Boot:** 1209/2308 **Towing:** 2550 **Seats**: 5 **Doors:** 5

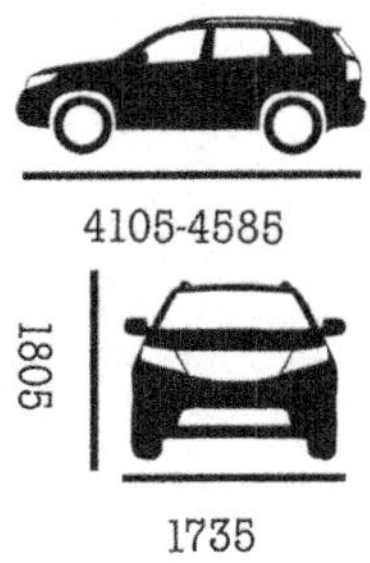

Ford Maverick 1993 - 1997

Also a Nissan Terrano, with a Primera interior, but the Ford badged version didn't do as well because of Oriental price/warranty cleverness. 3 and 5 door, as a standard and GLX model with alloys and wash/wipes. '95 chrome grille and some more kit. Engines revised in '96 before they were knocked off the price list. Wheel bearings and gearbox woes.
Engines: 2.4 • 2.7D **MPG**: 23-28.6 **Gearbox:** Man **Weight:** 1720 **Boot:**115/1900 **Towing:** 2500 **Seats**: 5 **Doors:** 3/5

Honda CR-V 1997 - 2001

A reinvention of the traditional estate car. Impressive standard kit from day one includes the usuals such as electric windows, air conditioning, sunroof and mirrors. Plus there was stuff to keep an adventurer happy such as a removable picnic table, an accessory socket in the boot and a split rear tailgate. The petrol 2.0 VTEC engine is all you need and it won't break down. Wear and tear and rust.

Engines: 2.0 **CO2:** 229g/km **MPG**: 29.1 **Gearbox:** Man/Auto **Boot:** 445/668 **Towing:** 1500 **Seats**: 5 **Doors:** 5

Honda HR-V 1999 - 2005

The world had no idea that it needed a 4 x 4 three door estate styled like a space shuttle. The world was also offered a two wheel drive SE version for a while. Five doors arrived in 2000 which eventually became the sole flavour and there was a complete range which was freshened up in 2001. Very likable, fun and practical. Suffers from rust and neglect.

Engines: 1.6 **CO2:** 201-205g/km **MPG**: 32.8-33.6 **Gearbox:** Man/Auto **Boot:** 285/603 **Towing:** 1200 **Seats**: 4 **Doors:** 5

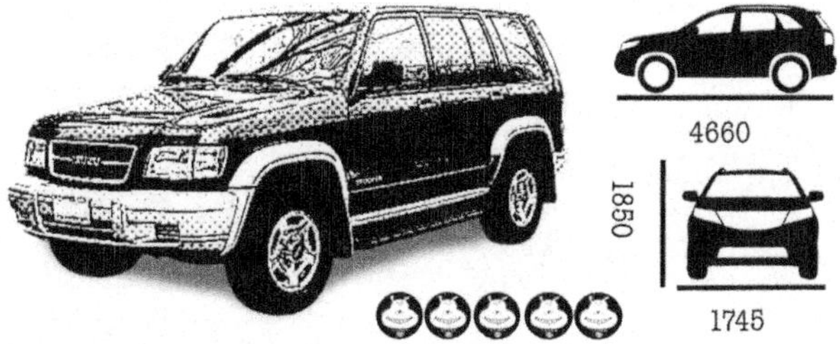

Isuzu Trooper 1992 - 2003

Called the Big Horn back home, third generation from '98 with a V6 petrol or new common rail diesel power. Loved by farmers and builders who just wanted to work and tow. Clever 4×4 system which constantly changed power delivery to the front or rear wheels. Not sophisticated but really tough. Legendary. May have been worked to death...

Engines: 3.5 • 3.0D **MPG**: 20.6-26.9 **Gearbox:** Man/Auto **Boot:** 909/1730 **Towing:** 3300-3500 **Seats**: 5 **Doors:** 3/5

Jeep Cherokee 1993 - 2001

Proper ancient off roader that copes with the rough and can tow all day long. Go for Limited specs that at least give the passenger an airbag and the driver gets ABS. Not massive inside but adequate. 2.5 Turbo diesel noisy and the 4.0 petrols not economical but pretty reliable and tough overall. Rust, worn suspension, tired smokey engines and gearboxes.

Engines: 2.5 • 4.0 • 2.5D **CO2:** 296g/km **MPG**: 18.2-31.1 **Gearbox:** Man/Auto **Boot:** 412/1265 **Towing:** 3150 **Seats**: 5

Jeep Grand Cherokee 1999 - 2004

Yank take on the whole LuxoSUV market that they invented. New generation Grand had just the one smaller V8 petrol. 3.1 Turbo diesel replaced by the better 2.7 CRD. Clever touches like the electrical retractable split tailgate glass. Dated but useful. Automatic gearbox, electrical niggles like locks and windows, general post towing scruffiness.
Engines: 4.0 • 4.7 • 2.7D **CO2:** 255-380g/km **MPG**: 17.7-29.1 **Gearbox:** Auto **Boot:** 1110/2059 **Towing:** 3360-3500

Jeep Grand Cherokee 1994 - 1999

Luxury SUV motoring from the originators. Massive 5.2 V8 left hooker only. These had ABS and airbags. Limiteds have everything as standard, revamp in '96 with the Laredo and top spec Orvis. Super comfy, tons of space and better than a Rangie. Check automatic gearbox, electrical niggles like locks and windows, general post towing scruffiness.
Engines: 4.0 • 4.7 • 5.2 • 2.5D • 3.1D**MPG**: 17.2-24.1 **Gearbox:** Auto/Man **Boot:** 1136/2254 **Towing:** 2265 Seats

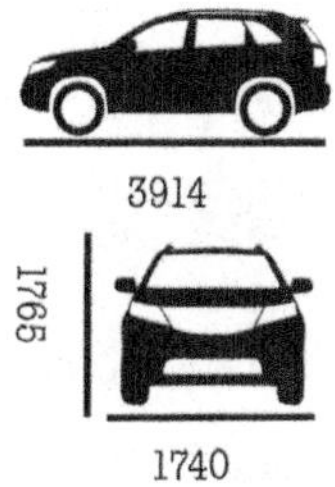

Jeep Wrangler 1993 - 2006

The 24 carat classic. Two or three doors, hard or soft top Proper trad styling from 1996 with the TJ when it got airbags and the Sahara model had a removable hardtop. Flappy plastic side windows, but really not worth bothering with. Should be used and enjoyed as a convertible 4 x 4. Back to basics tough Banger for the end of most days.

Engines: 4.0 **CO2:** 302g/km **MPG**: 22.1 **Gearbox:** Auto **Boot:** 320/1577 **Towing:** 1001-2000 **Seats**: 2/3 **Doors:** 2

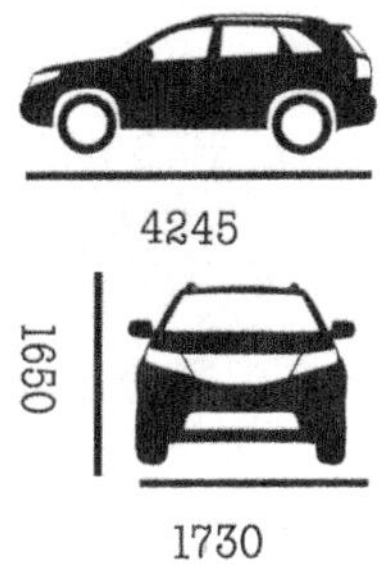

Kia Sportage 1995 - 2003

Ahead of the curve, well finished, handy sized SUV action. A Korean, built in Germany by Karmann, so decently finished. Just the one 2.0 petrol engine. Loads of different specifications over the years and restyled a bit in 2000. It is a no more than a slightly more capable estate car, but it is reliable. Wear +tear, rust and general old age and overuse.

Engines: 2.0 **CO2:** 275g/km **MPG**: 25.7 **Gearbox:** Man/Auto **Boot:** 483/1319 **Towing:** 1250 **Seats**: 5 **Doors:** 5

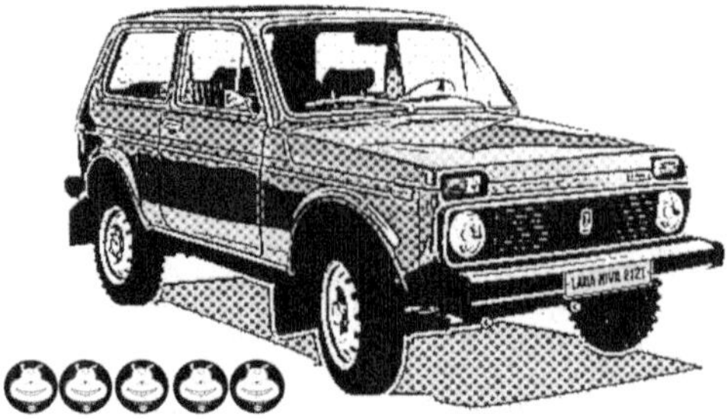

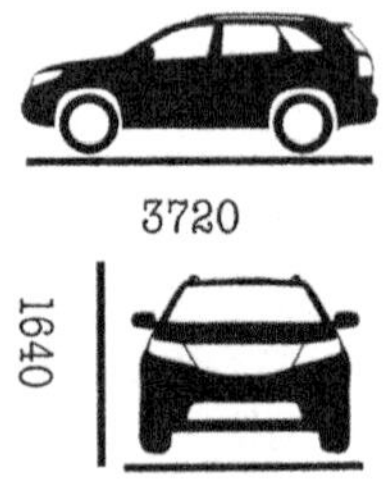

Lada Niva 1983 - 1998

A real word Defender that is rugged, excellent off road and properly basic. Cheap at the time and therefore expendable. Recently became cool despite being fairly uncomfortable, like a real Defender. Cossacks had more kit, revised in '93 with rood rails, new engine in '95. Low rent model became the Hussar. Electrics and charging issues, rust of course,

Engines: 1.6 • 1.7 **MPG**:22.7-29.8 **Gearbox:** Man **Boot:** 420/780 **Towing:** 1000 **Seats**: 5 **Doors:** 3

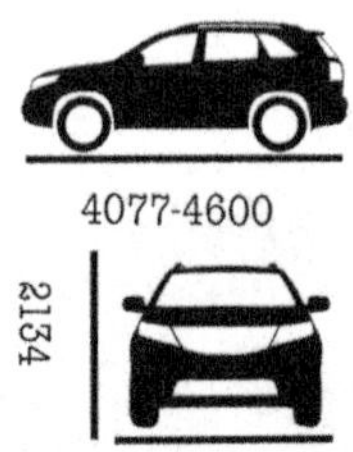

Land Rover Defender 1990 - 2012

Definitive working vehicle. It has the longest story ever told. The shape never changed much, just short 90 or long 110, with progressively better diesels. in 1999 it got the 5 cylinder unit from the Discovery. It remained an unsophisticated lorry. Never a cheap option. Rust, electrics. Once sorted very much a long termer.

Engines: 2.4D **MPG**: 28.2 **Gearbox:** Man **Boot:** 1510+ **Towing:** 3500 **Seats**: 2/7 **Doors:** 3/5

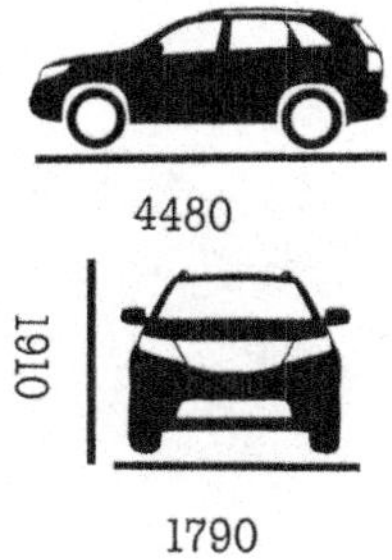

Land Rover Discovery 1989 - 1998

Range Rover with a different body on top, optional seven seats and some curious engine choices. The 300 TDi and V8s are the only ones worth considering as the tiny petrol dismally slow and troublesome. Spacious, refined and some decent XS and ES specs with twin sunroofs and alloys. Blown head gaskets, rot and electrical upsets. Better than II. **Engines:** 2.0 • 3.5 • 3.9 • 4.0 • 2.5D **MPG**:17-34.2 **Gearbox:** M/Auto **Boot:** 881-1699 **Towing:** 3500 **Seats**: 5/7 **Doors:** 3/5

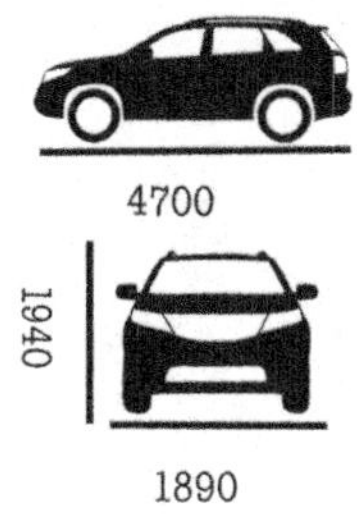

Land Rover Discovery II 1998 - 2004

Discovery II looked the same but only the tailgate shared with the cast iron original. Owners swear by and at them. Too much complicated technology meant the air suspension could fail and the electronics would have a mental breakdown. V8 petrol sounds great in between garage visits. Hard to find an upside. **Engines:** 4.0 • 2.5D **CO2:** 262-397g/km **MPG**:17-30.1 **Gearbox:** Man/Auto **Boot:** 881-1699 **Towing:** 3500 **Seats**: 5/7 **Doors:** 5

Land Rover Freelander 1997 - 2003

The first truly new Land Rover in a generation as a pint sized off roader. Three door Softback/Hardback and a five door estate petrols and diesels, but beware head gasket issues. Revised model in 2000 with a Td4 engine and specification shuffles. Keep an eye, or rather ear on the wheel bearings. Revised in 2003. **Engines:** 1.8•2.5•2.0D **MPG**: 22.7-37.2 **Gearbox:** Man/Auto **Boot:** 371/1311 **Towing:** 2000 **Seats**: 5 **Doors:** 5

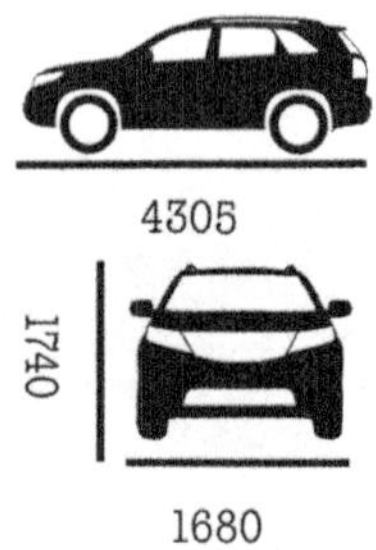

Mahindra Jeep 1990 - 1995

Based on the Willys CJ3B Jeep and built in India. However had a 4 speed gear box (as opposed to 3 speed of the original Willys), hanging pedals with a brake booster, a Peugeot Diesel engine. Dana 44 rear axle is tough and chassis made out of thicker material. Basic in the extreme and all the better for that. Contemporary owner feedback good.

Engines: 2.5D **MPG**: 25 **Gearbox:** Man **Boot:** 300/1500 **Load:** 700 **Weight:** 1450 **Seats**: 2/4 **Doors:** 2

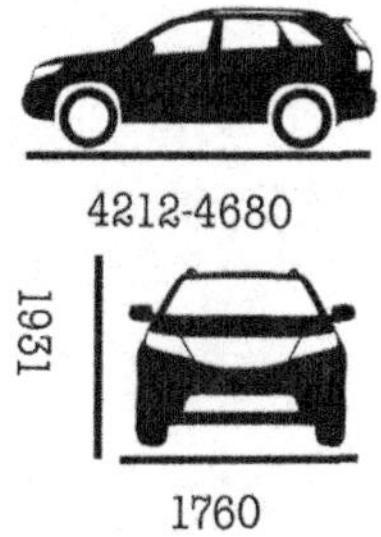

Mercedes G-Wagen 1982 - 1996

Old School Defender killer, but not remotely Range Rover comfy, just a boringly tough military go anywhere mobile. Uprated in '91 with 5 speed manual, central locking and alloys. Better than the Brits then. Became G-Class in '93 and gentrification process begins with air, driver's airbag, air and some walnut trim. Watch out for worn out ones.
Engines: 2.9 • 2.9D **MPG**: 16.2-20.3 **Gearbox:** Auto/Man **Boot:** 530/900 **Towing:** 750-3500 **Seats**: 5 **Doors:** 3/5

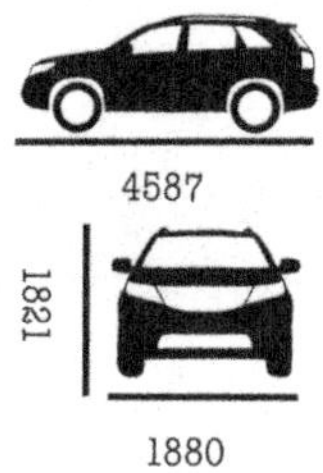

Mercedes M-Class 1998 - 2004

Yet another huge 4×4, but not a great time for quality control. That was sort of addressed in 2000 with a range revamp and then again in 2001, but lots of problems with the early examples and many are best left to rot and expire where they are parked. Decent petrol and diesel engines and optional third row of seats. Very scruffy ones around.
Engines:3.2 • 5.0 • 5.4 • 2.7D**MPG**: 19.6-30.0 **Gearbox:** Auto/Man **Boot:** 1061/2020 **Towing:** 2650-3365 **Seats/Doors**: 5

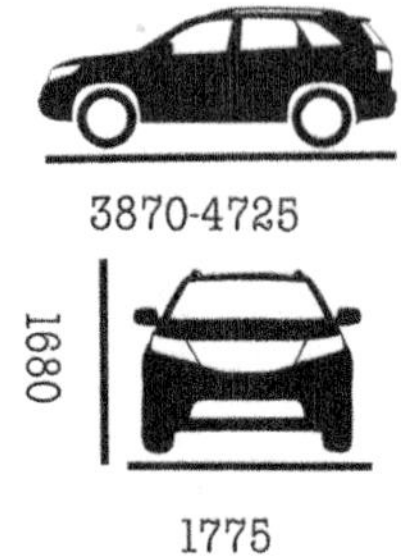

Mitsubishi Shogun 1983 - 2000

In case a buyer didn't want to break down all the time in a Range Rover here was the realistic on/off road answer in the shape of Shogun. Great on/off road and good for towing. Diamond Pack has most kit inc ABS. Short or long models with row three and 7 seats. Not much goes wrong, just fluid leaks and worn out parts. Lots of imports with big mirrors.
Engines: 3.0•3.5•2.5D•2.8D **MPG**: 19.5-25.8 **Gearbox:** M/Auto **Boot:** 1080-2050 **Towing:** 3000 **Seats**: 5/7 **Doors:** 3/5

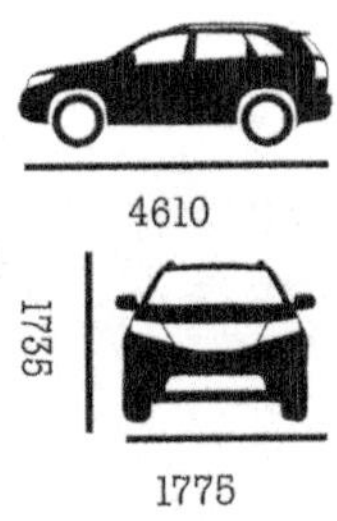

Mitzi Challenger Shogun Sport 1998 - 2006

Pick-up based Shogun, also known as a Challenger. There are compromises with no back seat tricks to pack stuff in, or fold anything flat. It is very basic business transportation. There's a decent 2.5 Diesel and thirsty 3.0 V6. Proper worker so many are worn out, or being sold because there are big bills to fix suspension or engines.
Engines: 3.0•2.5D **CO2:** 266-294g/km **MPG**: 22.6-27.2 **Gearbox:** Man/Auto **Boot:** 500/1590 **Towing:** 3100 **Seats**: 5

Nissan Patrol 1992 - 1998

Less comfy Shogun, but cheaper, ditto for Land Cruiser and Rangie, a big tough old unit that is best of down on the farm. Two sizes and the five door will have 7 seats. Parts can be a problem as there were never many around, not great mpg even on the diesels. Specification basic except on the SE and the later '90s have airbags. Neglected ones best left to rot. **Engines:** 3.0•4.2•2.8D•4.2D **MPG**: 17.5-26.5 **Gearbox:** Man/Auto **Boot:** 623/2170 **Towing:** 2800 **Seats**: 5/7

Nissan Patrol 1998 - 2009

Like a Defender, Jeep or Landcruiser, apparently around since the dawn of time. Big old bus and identified as GR models with three (until 2001) or five doors. Restyled in 2000 and in 2004 with thumpingly good diesel engines. Always better value than the more fashionable alternatives, yet these are rock solid. Watch out for overuse and abuse. **Engines:** 3.0D **CO2:** 288g/km **MPG**: 26.2 **Gearbox:** Man/Auto **Boot:** 183/2226 **Towing:** 3500 **Seats**: 5/7

Nissan Terrano 1993 - 1999

More popular and better priced than the Ford version which faded away whilst this was the family friendly SUV which predates the Cashcow. Reliable very obviously and the diesel is great. Revised in '96 and the TDi a big improvement, restyled a bit, SE Touring is the 7 seat one, lx is low spec. 3 Door. Most neglected some electrical and suspension issues.

Engines: 2.4•2.7D **MPG**: 23.7-29.1 **Gearbox:** Man **Boot:** 750/1600 **Towing:** 2800 **Seats**: 5/7 **Doors:** 3/5

Nissan Terrano II 1999 - 2005

Also available as a Ford Maverick, but that is just badge this is a proper Datsun. Three and five door versions and a range of diesel and petrol engines. Facelifted in 2000 but what you have is a pretty ordinary compact 4 x 4 was OK for the time (central locking, electric sunroof). At least it is resistant to breaking down. Wear and tear checks.

Engines: 2.4•2.7D **MPG**: 23.0-28.8 **Gearbox:** Man/Auto **Boot:** 1642 **Towing:** 2800 **Seats**: 5/7 **Doors:** 3/5

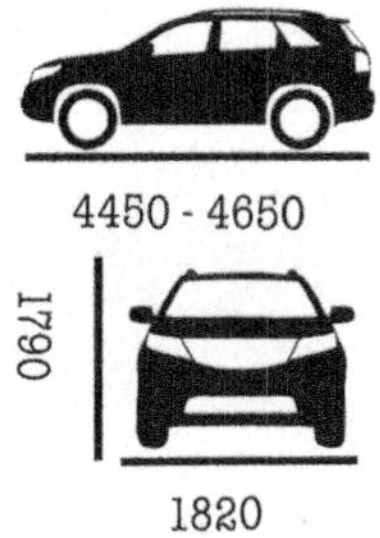

Range Rover 1989 - 1996

Luxury SUV defined. Quite old by the '90s and basic one badged as Classic from '94 with 'soft' dashboard. There's the rear legroom friendly LSE and most would want a Vogue SE. Incredible on and off road, but increasingly alternative to luxosaloons Still breaks down though, but what a classy way to do it. Oil leaks, rotten tailgate, shot suspension.

Engines: 3.9•4.2•2.5D **MPG**: 17.8-27.2 **Gearbox:** Auto **Boot:** 1020/2000 **Towing:** 4000 **Seats/Doors**: 5

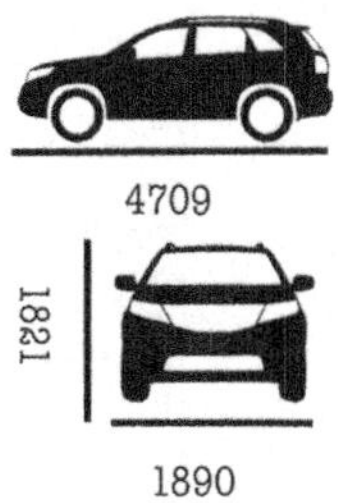

Range Rover 1994 - 2002

Powered by V8 petrols and 2.5TD diesel, the latter boosted for output in 1999. Cosmetic tidy up in 2000. Still flawed when it came to the basics, like reliability. Neglected and older ones are a liability, not the engines, mostly the electrics. Certainly does the on and off road better than anything else when it works but not worth the risk.

Engines: 4.0•4.6•2.5D **CO2:** 304-398g/km **MPG**: 17.2-24.9 **Gearbox:** Auto **Boot:** 524/1642 **Towing:** 4000 **Seats**: 5

Skoda Octavia 4×4 1998 - 2005

Working blokes Quattro, with the emphasis on being able to lug stuff and do a real world job rather than being lifestyle. Plus it has metal guide rails in the floor so you can slide stuff in and it is a metre wide. Diesel ECU failure, water pumps, air mass sensors and ignition coils across the engine range. Electrical failures in the cockpit.

Engines: 1.8 • 1.9D **CO2:**167- 211g/km **MPG**: 31.7 - 41.0 **Gearbox:** Man/Auto **Boot:** 560/1620 **Towing**: 1500 **Seats**: 5

Ssanyong Korando 1997 - 1998

Essentially an ugly Jeep and powered by a Mercedes five cylinder diesel which was good. Just the three doors, but is decently big inside with a good boot. This is basic 4×4 stuff which isn't bought to pose, but tow and work really hard which is not great when buying used, so rust and worn out mechanicals are the only concerns. Lived on as a Daewoo...

Engines: 2.3 • 2.9D **MPG**: 27.7-28.8 **Gearbox:** Man/Auto **Boot:** 479/1464 **Towing:** 2800 **Seats**: 5 **Doors:** 5

Ssangyong Musso 1995 - 1998

A quirky Korean which is tough as an everyday tow car with a big boot and more equipment than before. Mercedes related petrol engine and a turbo diesel are both pretty unburstable. Strictly five seats which is fine, just look out for rot and neglect. Fluid leaks and minor electricals, workhorse credibility with this one.

Engines: 2.3 • 3.2 • 2.9D **MPG**: 23.3-30.7 **Gearbox:** Man/Auto **Boot:** 1119-1920 **Towing:** 2800 **Seats**: 5 **Doors:** 5

Subaru Forester 1997 - 2003

Hard nut version of the Legacy and Impreza. Extra ride height, but mostly a brutal estate car with a 2.0 or from '98 Turbo petrol power. All Weather Pack installed air conditioning, contradictory sunroof and heated seats. Facelift from 2000 and apparently the rear seats are roomier. Everything costs a fortune to fix.

Engines: 2.0 **CO2:** 214-235g/km **MPG**: 28.5-31.4 **Gearbox:** Man/Auto **Boot:** 408/1526 **Towing:** 2555 **Seats**: 5 **Doors:** 5

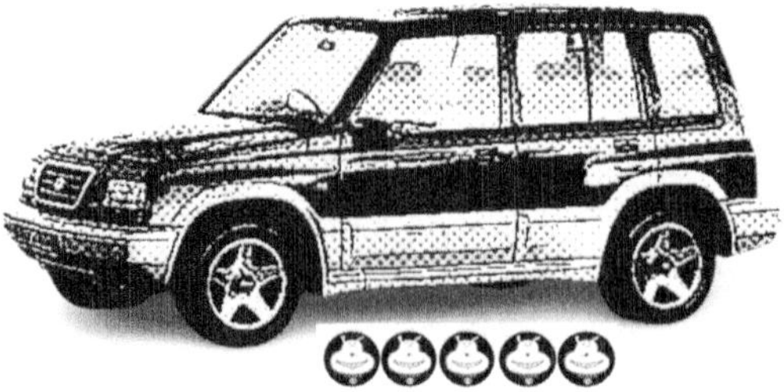

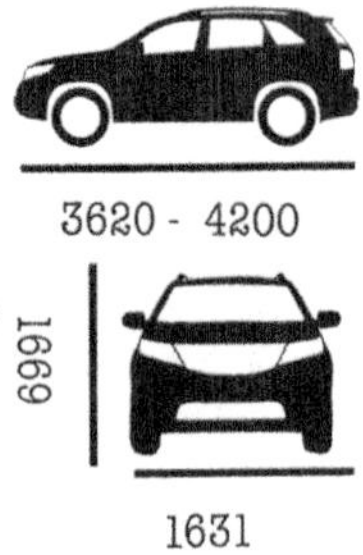

Suzuki Vitara 1988 - 1998

The vehicle responsible for the adoption of high rise family SUV of choice and what a great bit of kit. Tough, practical, stylish and usefully compact. Lots of models, this is the people's Landcruiser and potentially just as unbreakable. Revisions in '96 were style, not substance. Just minor electrical issues, look for coolant and oil leaks.

Engines: 1.6•2.0•2.0D **MPG**: 26.2 - 34.4 **Gearbox:** Man/Auto **Boot:** 202/845 **Towing:** 450 **Seats**: 5 **Doors:** 3/5

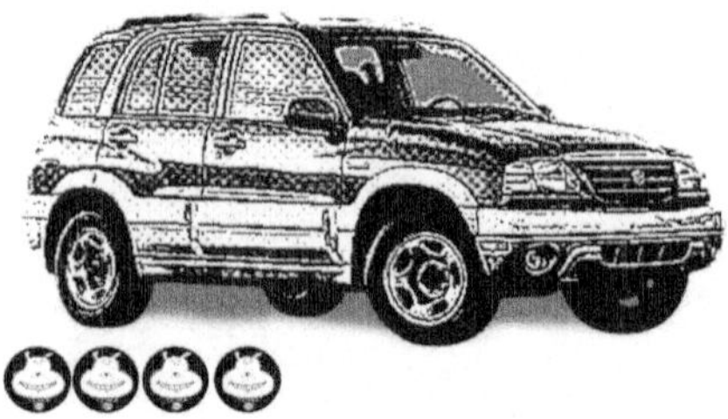

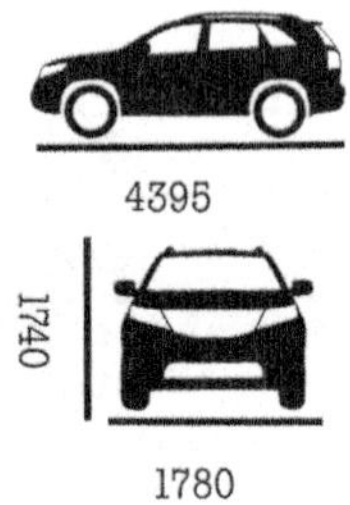

Suzuki Grand Vitara 1998 - 2005

A proper job when it comes to actually going off road, which should not be a surprise as it has an old fashioned chassis. Three and five door models, plus a soft top, so there is something for everyone there and the bigger XL7 in 2001 meant you could have more seats. Reliable, just minor electrical issues, look for coolant and oil leaks.

Engines: 1.6•2.0•2.5•2.7•2.0D **MPG**: 26.2 **Gearbox:** Man/Auto **Boot:** 572/984 **Towing:** 2000 **Seats**: 5/7 **Doors:** 3/5

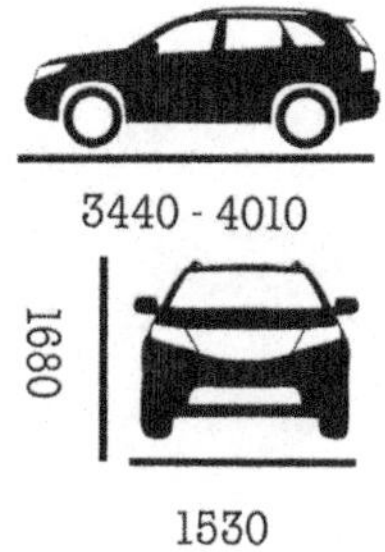

Suzuki SJ413 Samurai 1990 - 1996

Basic pint sized Jeep which excels as a holiday rental but will do so much more. A tough little nut with unbreakable engine, but Spanish built body as flimsy as you would expect in our climate. There's a hardtop, soft top and a longer one. Very underrated overall, poor on road but brilliant off for its size. Smokey engine, noisy suspension and rot.

Engines: 1.3 **MPG**: 33.2 **Gearbox:** Man **Boot:** 100ish **Towing:** 1100 **Seats**: 4 **Doors:** 3

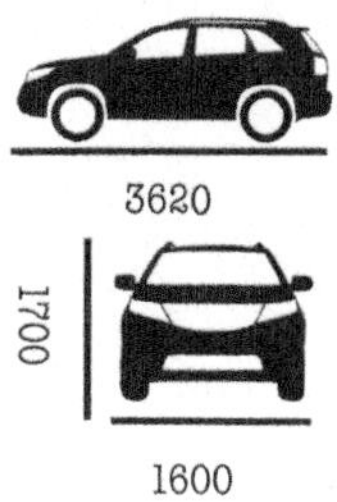

Suzuki Jimny 1998 - 2005

Pint sized Jeepney thing that is brilliant off road. Hard wearing with proper hose down plastic interior. Convertible model is a huge amount of fun, if that roof is missing or damaged, difficult and expensive to fix. The 1.3 petrol engine is keen and relatively simple to look after, just needing a cam belt every 70K miles. Can be killed by rust of course.

Engines: 1.3 **CO2:** 174g/km **MPG**: 38.7 **Gearbox:** Man/Auto **Boot:** 113/286 **Towing:** 450 **Seats**: 4 **Doors:** 3

Suzuki X90 1996 - 1998

Have I done the Barbie and Ken gag? Here's a targa that was available in two or four wheel drive. A soft Jimny that has no real purpose except as a 'hot rock' holiday rental. Looked stupid next to a RAV-4, but was well equipped, the four wheel drive is decent enough and hopefully being a looked after Suzuki, it won't breakdown.

Engines: 1.6 **MPG**: 36 **Gearbox:** Man **Boot:** 237
Weight: 1100 **Towing:** n/a **Seats**: 2 **Doors:** 2

Tata Gurkha 1992 - 2000

Basic utility vehicle was something of a Bitza with a decent Peugeot diesel engine, and switchgear from a mangy Austin Montego. There are two and four wheel drive models and some Range Rover rivaling spec with electric windows and a stereo radio-cassette. Owners seemed to like them. Good at being a farmyard hack, less happy driving into town.

Engines: 2.0D **MPG**: 31 **Speed:** 87 **Boot:** **Fuel Tank**: 11 gallons **Towing:** 3110 unbraked **Seats**: 4 **Doors:** 3

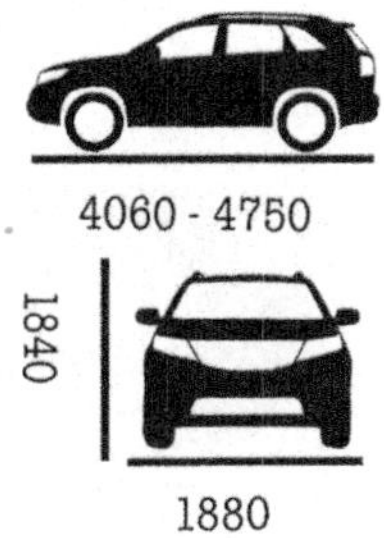

Toyota Land Cruiser 1988 - 1997

Legendcruiser. Not enough space to detail the pre history, effectively the 80 Series from 1990 marks the greatest 4×4 known to humankind, V8s petrols are great but the diesels always the better buys. Never cheap and rightly so. Superb at towing, lugging and just going where others struggle. Lasts forever with care and prices remain remarkably firm.

Engines: 4.2•4.5•4.2D **MPG**: 20-23.5 **Gearbox:** Man/Auto **Boot:** 275/1100 **Towing:** 3500 **Seats**: 5/7 **Doors:** 5

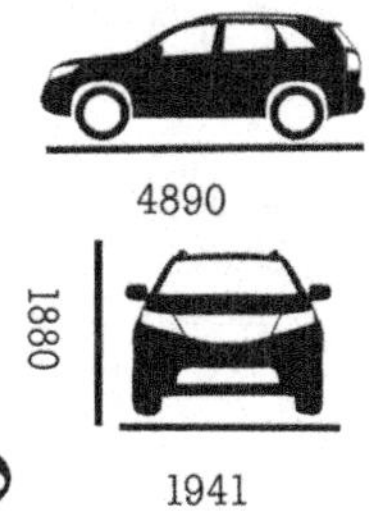

Toyota Land Cruiser /Amazon 1988 - 2008

The 100 series in 1998 continued where the old one left off just a bit smoother and luxuryfied. VX V8s petrols are fine, but the diesels always the better buys. Never cheap and rightly so. Great at towing, lugging and just going where others struggle. Won't break down if looked after. Lasts forever, no such thing as a cheap one of course.

Engines: 4.7•4.2D **MPG**: 17.3-25.4 **Gearbox:** Man/Auto **Boot:** 252/1114 **Towing:** 3500 **Seats**: 7 **Doors:** 5

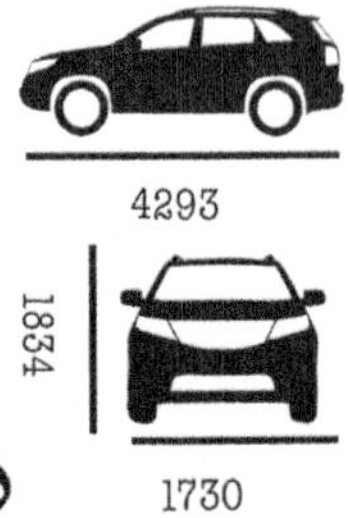

Toyota Land Cruiser Colorado 1996 - 2002

The so-called light duty six cylinder models or LJ90. There's a 3.0 diesel and 3.4 petrol. Everything that applies to the Amazon can be directed at the lickle school run one. Facelift in 1999, air conditioning and ABS became standard, plus different specs ending in X. Eight seaters plus power sockets. Check for leaks and damage. That's it.

Engines: 3.4•3.0D **CO2:** 253-341g/km **MPG**: 20-29.7 **Gearbox:** Auto **Boot:** 169/1100 **Towing:** 2800 **Seats**: 5/8

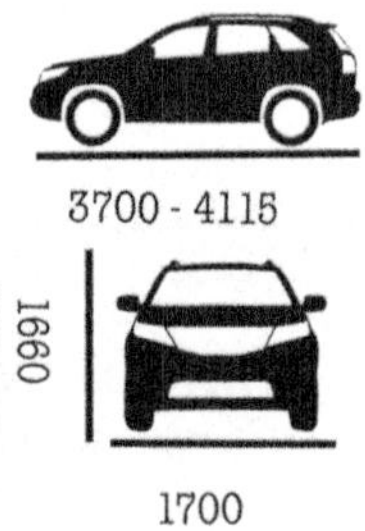

Toyota RAV4 1994- 2000

The original fun, funky thing a hot hatch in a perky compact SUV package. 5 door GX from '95, passenger airbag from '96, well equipped VX with air con in '97. facelift for '98. Perfect on road, not an offroader, you would get a Landcruiser for that. This was the future for a while. Few issues, just multi-owner servicing neglect.

Engines: 2.0 **MPG**: 30.1 **Gearbox:** Man/Auto **Boot:** 173/520 **Weight**: 1150 **Towing:** 1200 **Doors**: 3/5 **Seats**: 5

Toyota 4-Runner 1993- 1996

What grown ups bought instead of a fragile Discovery. Workaday image, but in real life rather comfortable and great to drive. Loadspace a bit awkward and not the biggest inside considering its size. Proper dual range transfer box and selectable four wheel drive. Nothing drops off or goes wrong unless it is being retired from the farm.

Engines: 3.0 • 3.0D • **MPG**:20.1-25.9 **Gearbox:** Man
Boot: 491/1705 **Towing:** 1500 **Seats**: 5

Vauxhall Monterey 1994 - 1999

Correct, this is an Isuzu Trooper with a Griffin badge, which should be a very good thing, buyers didn't flock to this dull Tonka. Short or long, it delivers the same levels of comfort and on/off road ability as the Tonka Trooper. Twin airbags from '97, seven seat LTD is the practical AntiPoshRover choice. Full fat Diamond Packs. Reliable. Run into the dirt.

Engines: 3.2 • 3.1D **MPG**: 19.3-25.1 **Gearbox:** Man/Auto
Boot: 1464/2548 **Towing:** 2600 **Seats**: 5 **Doors:** 3/5

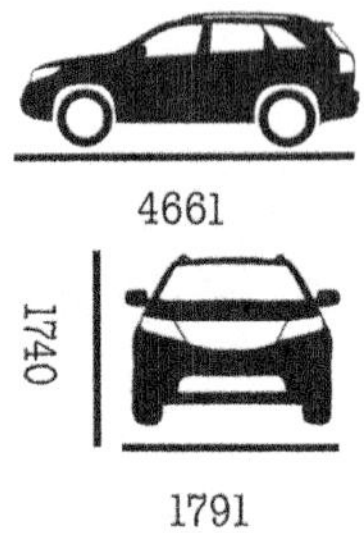

Vauxhall Frontera 1991 - 2003

The Isuzu Trooper connection didn't help much and built in Luton. Aimed at the high street rather than hard work. Revamped in '98 with new engines and styling which did not make it any more interesting. Part time four wheel drive, but essentially this is a big dumb vehicle with not much real charm. Towing damage, worn suspension, leaky engines.

Engines: 2.2•3.2•2.2D **MPG**: 22.4-31.0 **Gearbox:** Man/Auto **Boot:** 510/1784 **Towing:** 2400-2800 **Seats**: 5

BangerExecs - Management Motoring

The company car park was well established by the '90s and the steady transfer of of middle management preference from the mainstream to the premium manufacturers. If you come across a base specification anyexecmodel from this era, it was private sale. Fleet managers insisted on alloys, sunroofs and whatever made it SE to preserve the future resale value for their company.

Both BMW and Mercedes were a given, while Audi were powerfully making up the German triumvirate of power marques to be driving. In addition there is the arrival of Lexus as created in a laboratory by Toyota. The Mazda Xedos 9 did not strike a chord with anyone, see also the Mitsubishi Galant and Nissan QX. Oh and getting back to Toyota, their Camry, big in the USA, but never the UK.

This was the last gasp for the old school mainstream posho motors from Ford, the Granada and Scorpio which were a bit tired and outclassed. Vauxhall's Carlton and Senator still seemed relevant whilst the Rover 800 belonged in the 1980s. That Rover 75 looked as though it belonged in 1958, but actually it was a comfy old thing designed for the recently retired area manager.

You have to feel sorry for the Fiat Group who were still gamely trying to flog Alfa Romeos whilst we were enjoying the last days of Lancia being in showrooms before disappearing from Britain forever. Then there was the Fiat Croma, remember that?

Saabs and Subarus remained the weirdo company choices and for those who wanted to take a risk with a complicated Citroen XM, dull Renault 25 or one of those shonky old MGs, where a breakdown may have been just around the corner.

Alfa Romeo 75 1985 - 1992

Quirky looking at the time, but certainly purposeful with some OK to outstanding power units. Everyone agreed that the interior was obtuse but kept your attention. Putting the gearbox up the back meant it required concentration to shift. Ideal 50:50 weight distribution meant the V6 was especially lovely. Rust, electrics, everything costly. Great.

Engines: 1.8 • 2.0 • 2.5 • 3.0 **MPG**: 29-35 **Gearbox:** Man
Boot: 380 **Seats**: 5 **Doors:** 4

Alfa Romeo 155 1992 - 1998 

Sharp looking exec which was fabulously successful on the race track and first of the Fiat based Alfas, hiding a Tipo underneath. In 1995 more Fiat sourced engines and sharpened handling as it became a tad wider. The engines were pretty tough and it was the old issue of rot that finished many off prematurely. Worth having for rarity value alone.

Engines: 1.8 • 2.0 • 2.5 **MPG**: 27.6-34.5 **Gearbox:** Man
Boot: 525 **Seats**: 5 **Doors:** 4

Alfa Romeo 156 1997 - 2005

Inside and out this has always been a good looking car and great to drive. The level of standard equipment was pretty much kitchen sink and all models have climate control, central locking and electric windows. Failing oil pump and air flow sensors. Air con problems, Selespeed gearbox malfunctions.

Engines: 1.6•1.8•2.0•3.2•1.9D•2.4D **MPG**: 32.1-46.3 **Gearbox:** Man/Auto **Boot:** 374 **Seats**: 5 **Doors:** 4

Audi 80 1991 - 1996

Functional rather than fun and reshaped for a new decade, built to a sky high quality level. Came with mostly zero spec and small sluggish engines. A bit tight inside and the 2.0E best entry model and the V6 for smooth progress. Diesels are great for pure low level economy. Tough and just become smoky, scruffy and overworked, being galvanised helped.

Engines: 1.6•2.0•2.3•2.8•1.9D **MPG**: 31.3-44.5 **Gearbox:** Man/Auto **Boot:** 453 **Doors:**4 **Seats**: 5

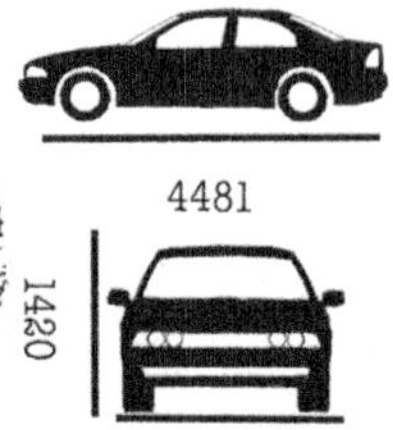

Audi A4 1995 - 2001

Default business saloon and from '99 a revamp tweaked the suspension and steering for the better. Rear seat passengers remained a bit cramped and endured the rock hard suspension. earlier models did have some mean specifications, especially below SE. It is the wear and tear issues of brakes and suspension, otherwise it is just service history. **Engines:** 1.6 • 1.8 • 2.4 • 2.6 • 1.9D • 2.5D **MPG**: 23.0-53.0 **Gearbox:** Man/Auto **Boot:** 439 **Seats**: 5

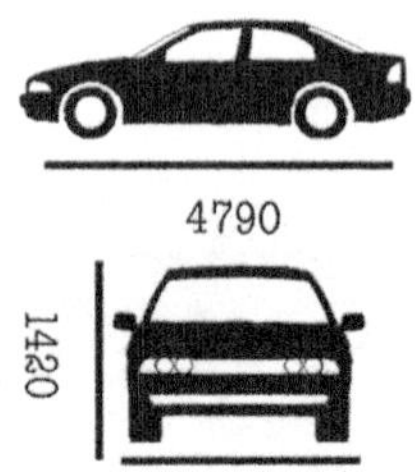

Audi 100 1991 - 1994

Slippery galvanised saloon car legend makes it into the '90s under its own power. Safe, comfortable and refined, with some great engines, from the TDI diesels to the smooth V6. There was even optional 6 speed gearboxes and Quattro all wheel drive. So can get complicated, smoky engines, noisy gearboxes and bearings as the miles mount up.
Engines: 2.0 • 2.2 • 2.3 • 2.6 • 2.8 • 2.5D **MPG**: 26.1-44.1
Gearbox: Man/Auto **Boot:** 510 **Seats**: 5

Audi A6 1994 - 1997

Gets a new paper size name but underneath was still a good old 100. hard to see the difference, just a bit better to drive. The usual colossal number of engine options, wonderful V6s and epic diesels that deliver a solid 45mpg. Quattros if you need them with lower suspension and top SE spec. Simpler ones are the solid Bangers for life.

Engines: 1.8 • 2.0 • 2.6 • 2.8 • 1.9D • 2.5D **MPG**: 26.3-47.9 **Gearbox:** Man/Auto **Boot:** 510 **Seats**: 5 **Doors:** 4

Audi A6 1997 - 2004

An all new safer and more spacious saloon with the usual ton of engine and drivetrain options. It never changed much. Best to avoid the Tiptronic clutchless manual on the grounds of complication. They all have twin airbags and climate control. Fuel pumps and air flow meters can fail, wheel bearings. **Engines**: 1.8 • 2.0 • 2.0 • 2.7 • 3.0 • 4.2 • 1.9D • 2.5D **MPG**: 21.7-49.6 **Gearbox:** Man/Auto **Boot:** 549 **Seats**: 5

BMW 3 Series 1983 - 1991

Legendary saloon slips into a new decade along with other variations, but deserves its place here as one of the best pre modern Bangers. Great 6 cylinder engines, the fours will plod on though. Late model uprated specs SEs with cross spokes and spoilers. Killed by rot, blown head gaskets, worn out suspension and just general neglect in the old days.
Engines: 1.6 • 1.8 • 2.0 • 2.5 **MPG**: **Gearbox:** Man/Auto **Boot:** 425 **Seats**: **Weight:** 1060 5 **Doors:** 4

BMW 3 Series 1991 - 1998

Not quite as E30 solid but probably slicker and certainly fine to own and drive, Everyone wanted one. Usual slant of good fours, great sixes and arrival of the impressive turbo diesels. Again, the 6-cylinder diesel was astounding. Revamped in '96 a bit, Sports, SE and Lux packs. Pretty reliable, but fluid leaks, iffy trim fragile, some electrics go 'bong'.
Engines: 1.6 • 1.8 • 2.0 • 2.5 • 2.8 • 1.8D • 2.5D **MPG**: 31.0-50.4 **Gearbox:** Man/Auto **Boot:** 435 **Seats**: 5 **Doors:** 4

BMW 3 Series 1998 - 2005

Prettiest 3 series? Good range of engines both petrol and diesel. Hard to go very wrong, just depends on what you can find and afford. Available in several exciting model flavours, but the saloon is almost the perfect small executive. Power loss and smoke, diesel issues, broken air mass meters and cracked inlet manifolds. Worn balljoints.

Engines: 1.8 • 2.0 • 2.2 • 2.5 • 3.0 • 1.8D • 2.0D • 3.0D
MPG: 31.0-50.4 **Gearbox:** Man/Auto **Boot:** 439

BMW 5 Series 1988 - 1996

Models goes up a few gears in comfort, quality and sheer presence, the premier Execubanger of the time. 4WD X model in 1992. V8s are great, but a big 6 is the decent compromise. 518 can't cope, 530i can. Plenty to go wrong if not cared for, electronics, suspension bushes, injectors on diesels, whilst there is smoke and gearbox faults.

Engines: 1.8 • 2.0 • 2.5 • 3.0 • 3.5 • 4.0 • 2.5D **MPG**: 22.3-39.6
Gearbox: Man/Auto **Boot:** 460 **Seats**: 5 **Doors:** 4

BMW 5 Series 1996 - 2003

The last of the fixable and greatest BMWs, (E39, anoraks). The diesels are wonderful especially the bigger sixes which offer thumping performance combined with regular 40mpg+ economy. Otherwise the petrol sixes will last forever, if not longer. Check electronics on old ones, suspension bushes. Diesels may need new injectors.

Engines: 2.0 • 2.2 • 2.5 • 3.0 • 3.5 • 4.0 • 2.5D • 3.0D **MPG**: 23.0-42.2 **Gearbox:** Man/Auto **Boot:** 460 **Seats**: 5 **Doors:** 4

Citroen XM 1989 - 2000

Still delivering unconventionality by the bucket load it also meant supreme comfort. The manuals are bit clunky, and 2.1 diesel agricultural but what a wonderful way to travel. Excellently space age interior and dash. Requires bravery to own with slipping auto gearboxes, broken clutches on the manuals and all sorts of fun with Hydractive suspension.

Engines: 2.0 • 3.0 • 2.1D • 2.5D **MPG**: 24.4-41.9 **Gearbox:** Man/Auto **Boot:** 435/1120 **Seats**: 5 **Doors:** 5

Ford Granada 1985 - 1994

Final shout for the Granny? That meant hatches and the posher saloon which arrived in 1990 especially in Scorpio trim (above Ghia and Ghia X) and with the full fat 2.9 24V. The turbo diesel is a bit of slug, 2.0 petrol best all round, lots of room, was easy and cheap to fix at the time. Loads of smoke, worn cranks, wheel bearings and soggy suspension.
Engines: 2.0 • 2.9 • 2.5D **MPG**: 24.8-35.6
Gearbox: Man/Auto **Boot:** 465 **Seats**: 5 **Doors:** 4/5

Ford Scorpio 1995 - 1999

Here is the Guppy one, designed to be different and last of the ExecuFords Six engines on offer, and loads of specs the Executive was the basic one, then Ghia, Ghia X and Ultima. All had useful 60/40 folding seat, metallic and music system. Late life minor restyle in '98. Unloved at the time, had minor electricals, e.g. windows, otherwise engines and suspension.
Engines: 2.0 • 2.3 • 2.9 • 2.5D **MPG**: 23.9-32.8 **Gearbox:** Man/ Auto **Boot:** 465 **Seats**: 5 **Doors:** 4/5

Fiat Croma 1986 - 1993

Centre section related to the Saab/Alfa/Lancia, variations on a 2.0 power plant, from normal to turbo. No one really wanted this Italian one, so cheap and neglected. The old thing was revised in 1991 with 'balancer' shafts so smoother and restyled slightly with marginally better interiors. Most dead, may have blowy turbos and worn out gearboxes.

Engines: 2.0 **MPG**: 34-39.3.1 **Gearbox:** Man/Auto **Weight**: 1142 **Boot:** 498 **Seats**: 5 **Doors:** 5

Jaguar S-Type 1999 - 2007

First of all it came with a 3.0 V6 and 4.0 V8. The V6 diesel arrived in 2004 following on from a mildish facelift so it made company car commonsense. Comfortable, easy to operate and pretty quick when required. Take advantage of its unlovedness. V8 timing chain tensioners. Rust, expensive suspension parts, electrics like windows.

Engines: 3.0•4.2•2.7D **CO2:** 179-259g/km **MPG**: 24.5-41.5 **Gearbox:** Man/Auto **Boot:** 400/810 **Seats**: 5 **Doors:** 4

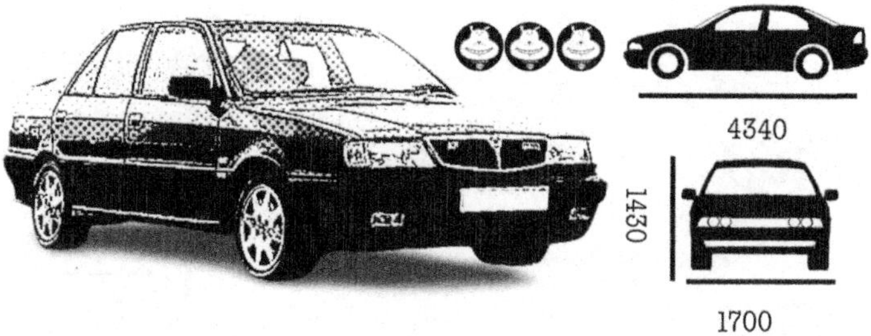

Lancia Dedra 1990 - 1994

Sort of Italian 3 Series, better value new and cheapo when used. Terrible name, certain amount of style, nippy turbo, but handling is something of a handful. Zero company car park appeal helped kill it. Just all the usual what to look fors with any '90s Fiat product that was pretending to be special. Ultimate retro cool saloon for just a few Lira.

Engines: 1.6 • 1.8 • 2.0 **0-60**: 7.2-13.4- **Speed:** 112-134
MPG: 32-34 **Gearbox:** Man **Boot:** 480 **Doors:** 4

Lancia Thema 1989 - 1994

Second generation is a sort of Italian 5 Series. It had new headlamps and 16v engines, developing up to 200bhp. A further subtle facelift followed in 1992. Blown head gaskets, rust of course, always needed proper care, seldom got it from owner two onwards and turbo is especially fragile. Worn gearbox and driveshafts just some of the problems.

Engines: 2.0 • 2.8 **0-60**: 10-13.4- **Speed:** 121-137
MPG: 30-34 **Gearbox:** Man **Boot:** 550/1000 **Doors:** 4

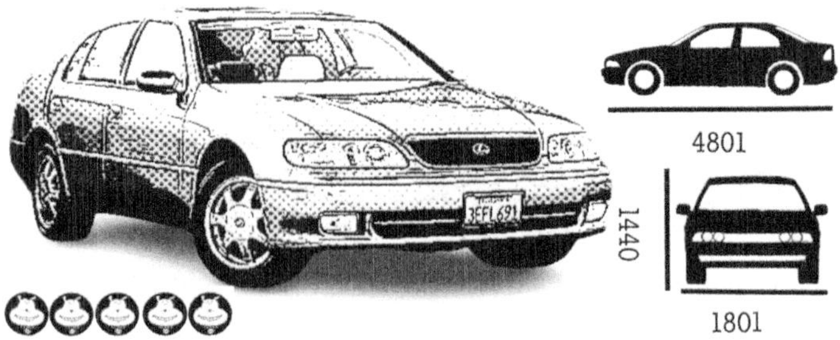

Lexus GS 1993 - 2004

What a mid size Jag should be, ultra reliable with the highest of technical technology. Very safe and roomy enough, most things are standard, except a diesel, they didn't exist back then. V8 or 6 cylinder. Check suspension and brakes. indeed all parts are pricey despite being a Toyota. Can be rusty.

Engines: 3.0•4.3 **CO2:** 281-296g/km **MPG**: 23-24.4
Gearbox: Auto **Boot:** 515 **Seats**: 5 **Doors:** 4

Lexus IS 1999 - 2005

Started life as a Toyota Altezza and that is reason enough to buy one. Copycat 3 Series saloon. Just the one 2.0 6 cylinder petrol engine until the IS300 arrived. SE has most fixtures and fittings and there is the LE too. Best not to buy a wide boy gold plated one. No issues, just wear and tear as usual to fix the brakes, suspension and any missing parts.

Engines: 2.0•3.0 **CO2:** 233-256g/km **MPG**: 26.2-29.1
Gearbox: Man/Auto **Boot:** 399 **Seats**: 5 **Doors:** 4

Mazda Xedos 9 1994 - 2001

Not desirable, but quality kit. Great engines, a smooth V6 then from '98 clever 'Miller Cycle' (smaller size, but behaves like something bigger) along with more anonymous restyle. Well put together. Good specs, SE had even more. Interestingly dull oddity. Rust is now an issue. Can look quite rough, it is tyres, exhaust and brakes that cost.

Engines: 2.3 **CO2:** 245g/km **MPG**: 28.2 **Gearbox:** Auto **Boot:** 416 **Seats**: 5 **Doors:** 4

Mercedes 190 1982 - 1992

Best Beige Banger ever. Covered colossal mileages in some style and from a zero spec starting point. Despite being a top Taxi, tight inside but solid. Autos and 2.0 best, 2.6 subtle, small diesel sluggish. Overall though the perfect specimen for proper Bangernomics. Most were driven to star ship mileages, so it was always worn out engines, suspension...

Engines: 1.8 • 2.0 • 2.5 • 2.6 • 2.0 • 2.5D **MPG**: 30.2-43.5 **Gearbox:** Man/Auto **Boot:** 410 **Seats**: 5 **Doors:** 4

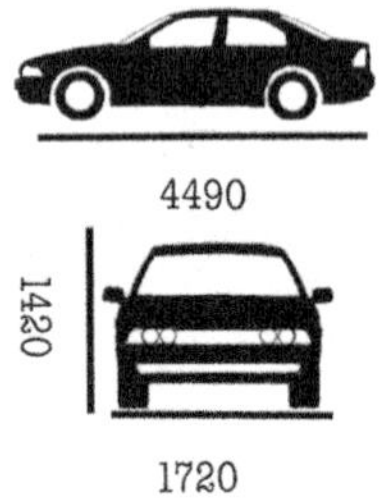

Mercedes C-Class 1993 - 2000

Halfbaked 190. Officially more wonderful, realistically less old school solid Merc more style than classy substance. 1997 revamp was considerable new engines and silly subtle styling things like dark tail lights, if that was your bag. The diesels are great of course. Still rusted though. Indeed rusty suspension parts, wipers cost a bomb to fix, engine oil leaks.
Engines: 1.8 • 2.0 • 2.3 • 2.4 • 2.8 • 2.2D • 2.5D **MPG**: 26.6-45.6
Gearbox: Man/Auto **Boot:** 430 **Seats**: 5 **Doors:** 4

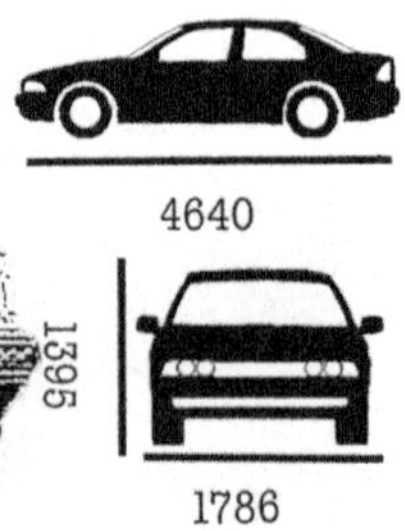

Mercedes E-Class W124 1984 - 1995

Premium E, another Taxi legend. Epically reliable, comfortable and nothing Merc has ever made since is as solid. Colossal number of engines. Mercedes added more safety overt the years so an airbag in '92 and a passenger one in '94. Massive mileages and neglect take their toll, the expensive bits are automatic gearboxes and rear axle.
Engines: 2.0 • 2.2 • 2.6 • 2.8 • 3.0 • 3.2 • 5.0 • 2.5D • 3.0D
MPG: 19.6-38.9 **Boot:** 465/665 **Seats**: 5

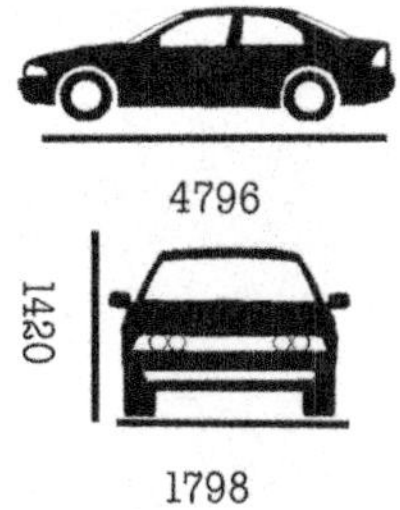

Mercedes E-Class 1995 - 2002

Es were good, but these were disappointingly rusty. New school saloons not as solid or special as the old W124. Usual diesels and petrols and substantially revamped in 1999, so better to buy those if you must. The common rail diesels are good and the petrols reliable. Oil leaks, rust to brake lines, suspension bushes, engine mountings. More rust.

Engines: 2.0 • 2.4 • 2.8 • 3.2 • 4.3 • 2.1D • 2.7D • 3.2D **MPG**: 25.0-45.6 **Gearbox:** Man/Auto **Boot:** 520 **Seats**: 5

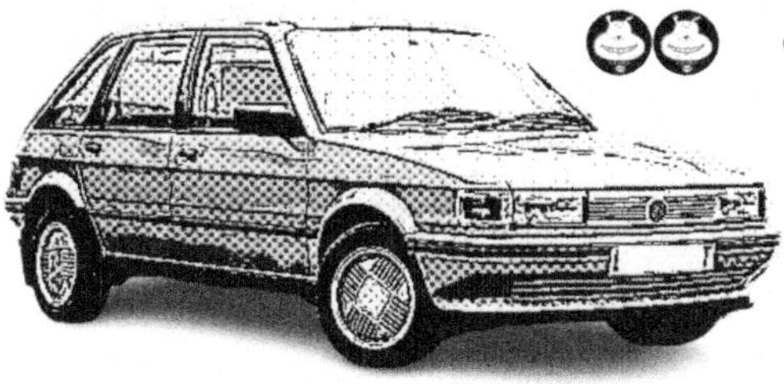

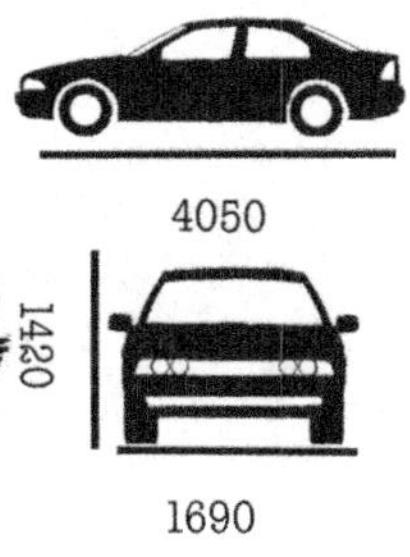

MG Maestro 1983 - 1994

By the '90s the Turbo had gone and then it was left to the hatchback to keep the MG brand going. The 2.0EFi did the honours and it was a dated old thing compared to the 200, but at least practical, even if the insurance premium was higher. Most left to fester before they became cool. So rotten sills blown head gasket and knackered suspension.

Engines: 2.0 **MPG**: 30 **Gearbox:** Man **Boot:** 304/1091 **Speed**: 115 **0-60**: 9 **Seats**: 5 **Doors:** 5

MG Montego 1985 - 1991

Just made it into the '90s as an outclassed saloon, but at least the Turbo was fun. The second generation ones were far better to drive and not the headstrong and untidy. Retuning the suspension was a revelation. Decent equipment levels by the end. In the end no one was bothered by them as they rusted and blew head gaskets.

Engines: 2.0 **MPG**: 30. **Top Speed**: 129 **0-60**: 7.5 **Gearbox:** Man **Boot:** 520 **Seats**: 5 **Doors:** 4

Mitsubishi Sigma 1991 - 1995

Japanese 5 series and largely forgotten. Had trick four wheel steering, electronically controlled suspension so super clever and complicated. Had leather and cruise. Revised a bit in '93. Got high intensity headlamps and uprated trim a year later and that was it. Highly reliable but needed pukka looking after. No image, meant no one cared.

Engines: 3.0 **MPG**: 24.7-25.6 **Gearbox:** Man/Auto **Boot:** 428 **Weight**: 1468 **Seats**: 5 **Doors:** 4

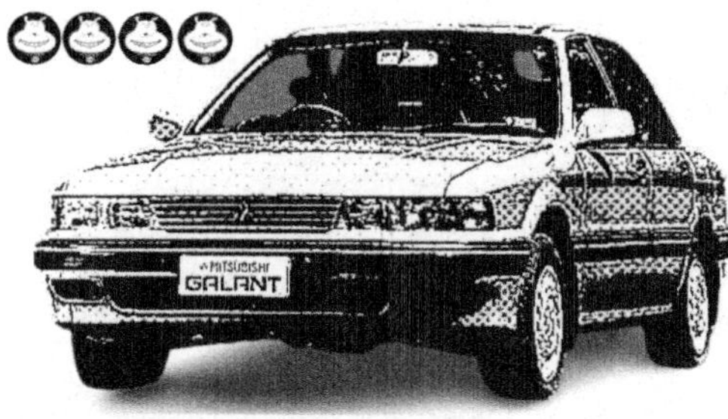

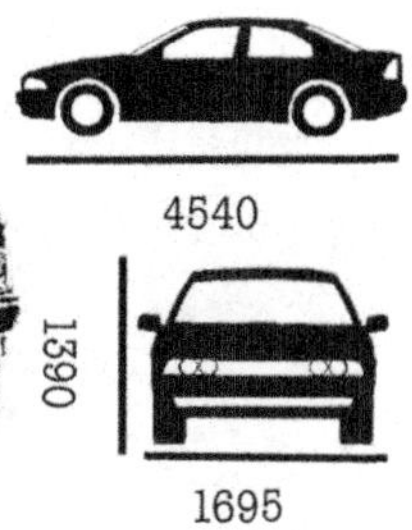

Mitsubishi Galant 1988- 1997

Here you go an old school saloon and a new fangled five door hatchback coupe. There was a new series with a split grille from 1991. Then in '95 a low spec 1.8 GLi joined the range. However, there was a 2000 GTi with four wheel drive and four wheel steering, plus cruise and electric mirrors. The simpler ones last as long as is required with proper fettling.

Engines: 1.8 • 2.0 • 2.0 V6 • 2.5 **MPG**: 26.3-36.9
Gearbox: Man/Auto **Boot:** 400 **Seats**: 5 **Doors:** 4/5

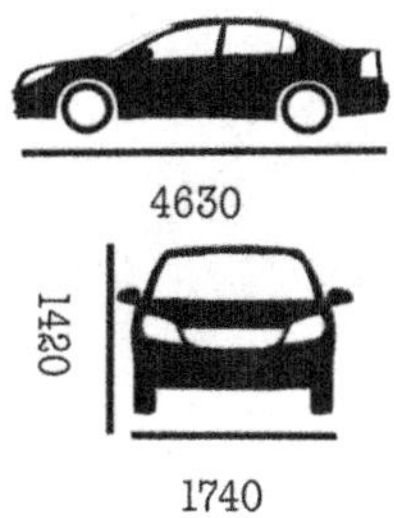

Mitsubishi Galant 1997 - 2003

Japan's car of the year '97 apparently. It featured a GDI engine, a gas direct injection unit which was jolly efficient even if Euro fuel wasn't always pure enough for it. Not a big hit, few around, classy and solid, proper alternative to one of those German saloons. Parts are very expensive when required. Check gearbox condition, brakes and suspension.

Engines: 2.0 • 2.4 • 2.4 **CO2:** 203-207g/km **MPG**: 29.7-33.2
Gearbox: Man/Auto **Boot:** 470 **Seats**: 5 **Doors:** 4

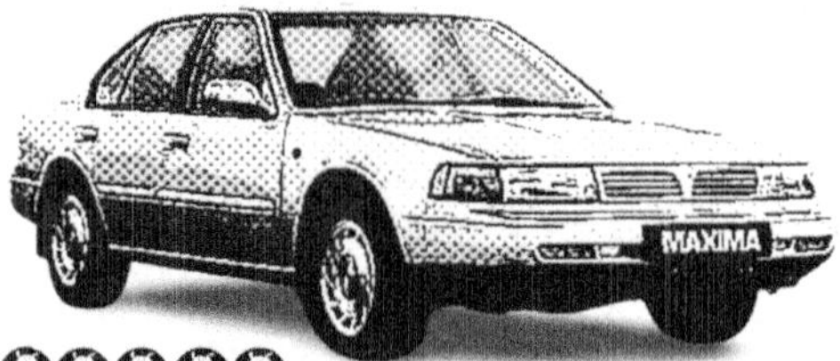

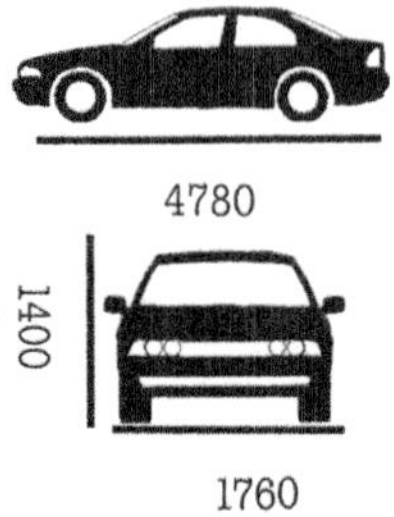

Nissan Maxima 1989 - 1994

Designed to take on the mid size Yank Tanks, in Murica it had the 3.0-V6 single-cam engine under the 'hood' is similar to the one in the 300ZX. Had ABS too and moquette upholstery, SEL with leather. Driver's airbag from '92 after a short break. Good luck tracking one down, they will last as long as an owner needs them to.

Engines: 3.0 **MPG**: 29 **Gearbox:** Auto **Boot:** 452
Speed: 137 **0-60:** 8.7 **Seats**: 5 **Doors:** 4

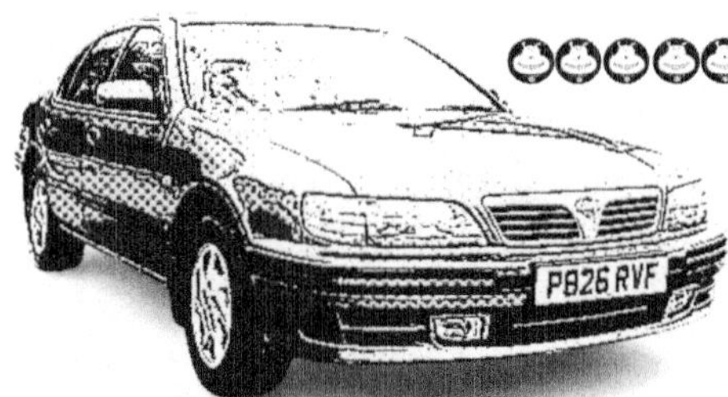

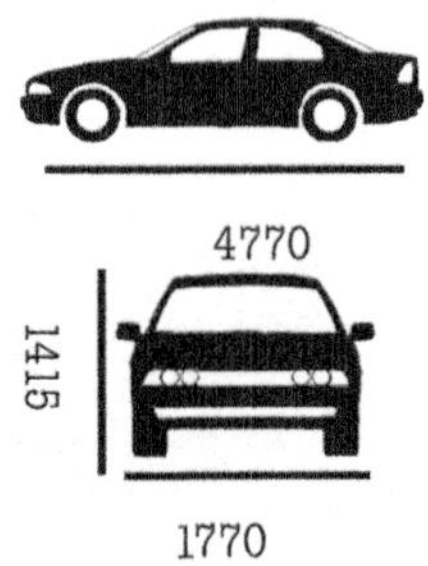

Nissan QX 1995 - 2000

Big saloons were never duller and anyway everyone else bought a Vauxhall Omega. Marginally larger than a Maxima plus there was a 2.0 V6 plus extra specification in the SEL with cruise, limited slip diff and electric seats. '97 upgrade with a new grille and other minor touches. Won't break down, but could sap your will to live.

Engines: 2.0•3.0 **MPG**: 26.4-28.5 **Gearbox:** Man/Auto
Weight: 2005 **Boot:** 440 **Seats**: 5 **Doors:** 4

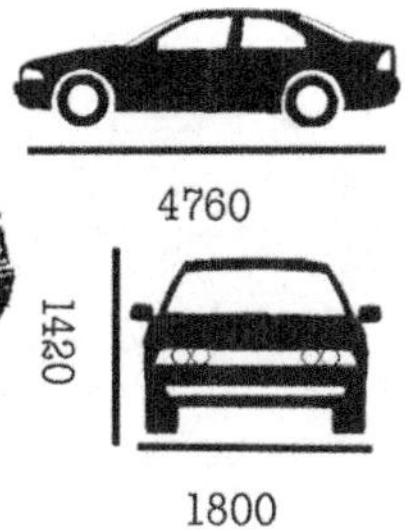

Peugeot 605 1990 - 1997

Here's a picture of depreciation which meant it became a taxi once retired from company car motorway duties. Otherwise best known for delivering comfort with a fabulously subtle ride. The 2.0 turbo great and most went for the diesel anyway. SV had cruise and automatic ride control, SVE delivered leather. Clutch slip and suspension wear.

Engines: 2.0 • 3.0 • 2.0D • 2.1D **MPG**: 24.4-41.9 **Gearbox:** Man/Auto **Boot:** 497 **Weight**: 1475 **Seats**: 5 **Doors:** 4

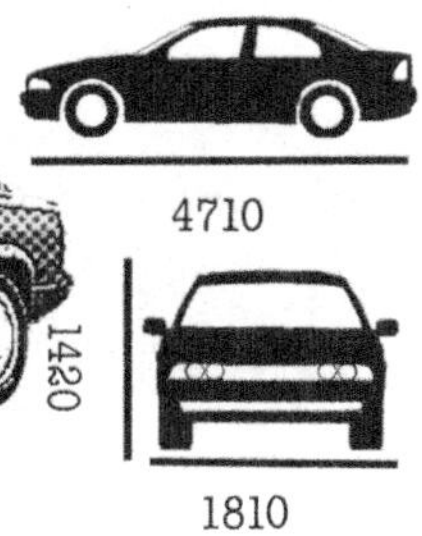

Renault 25 1984 - 1992

Francoluxury which means bags of space, comfort and a wonderfully smooth ride. No diesels, just a bunch of not especially frugal petrols plus late model Executives with lots of kit. Endearing giant hatch which was never cherished enough to be a longer term survivor. Electrics, head gaskets, overheating, auto gearbox failure, everything really.

Engines: 2.0 • 2.1 • 2.5 • 2.7 • 3.0 **MPG**: 19.2-32
Gearbox: Man/Auto **Boot:** 442 -1238 **Seats**: 5 **Doors:** 5

Renault Safrane 1992 - 1999

Big in Europe it lost out in the UK to the Germans, this big dumb hatch is refined and well equipped. Better built than a 25 and had a diesel so made some sense. Second generation in '96 Volvo 2.5 was used and subtle styling changes. Everything was an Executive at the end with top notch specs. Smoking small engines, dicky sunroof and electrics.
Engines: 2.0 • 2.2 • 2.5 • 3.0 • 2.5D **MPG**: 23.8-34.4 **Gearbox:** Man/Auto **Weight:** 1370 **Boot:** 481 **Seats**: 5 **Doors:** 5

Rover 75 1999 - 2005

A BMW retro-reimagining of what a retired colonel drove in 1958. Three petrols and diesel and carefully crafted trim levels. Avoid head gasket K-Series issues and uninspiring 2.0 V6 engines. The diesel is probably the pick of the bunch. Rotten sills and wheelarches, blown head gasket, diesel pump failure is major grief. Corroded suspension.
Engines: 1.8 • 2.5 • 2.0D **CO2:** 163-225g/km **MPG**: 29.2-48.8 **Gearbox:** Man/Auto **Boot:** 432 **Seats**: 5 **Doors:** 4

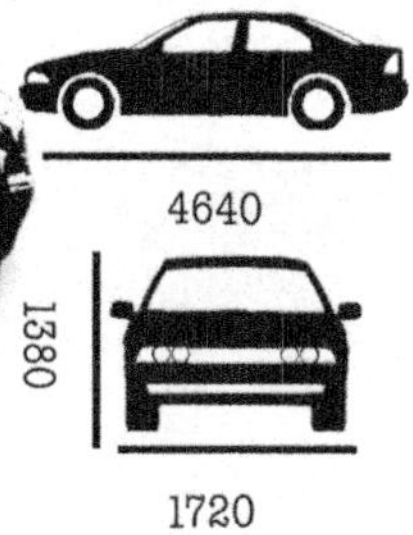

Rover 600 1993 - 1999

A proper Ronda, underneath it's an Accord with a mix of Honda + Rover propulsion. It certainly hit the buyer spot with a dab of walnut and chrome for the full trad image. '94 small improvements to Racks up some serious mileages and keeps going strongly. SD and LD are the top spec ones. Rust killed most from the early days, electricals and gearbox.

Engines: 1.8 • 2.0 • 2.3 • 2.0D **MPG**: 30.4-49.6
Gearbox: Man/Auto **Boot:** 430 **Seats**: 5 **Doors:** 4

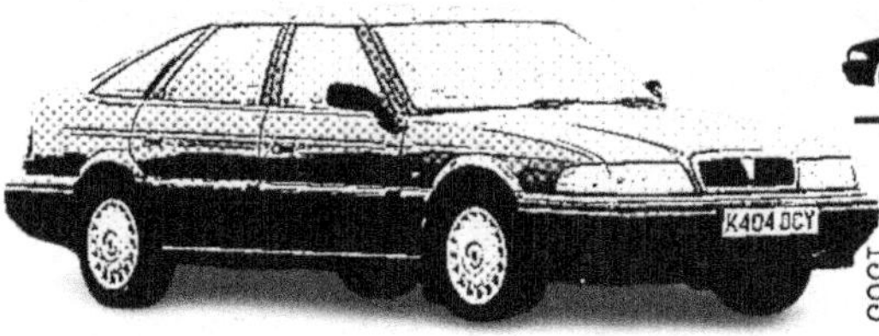

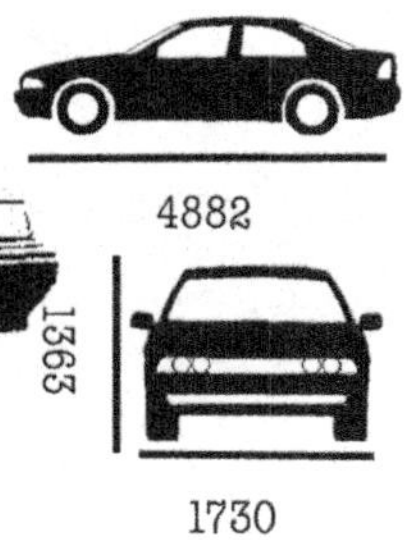

Rover 800 1991 - 1999

Revamped for a new decade with a new body and engine options, in particular the 2.0 Vitesse. Better, but not brilliant, only average overall. Updated spec in '94 with driver's airbag and passenger one in '96 when revised again. It should have died then, but stayed long enough to deliver broken suspension, sticking electric windows etc...

Engines 2.0 • 2.5 • 2.7 • 2.5D **MPG**: 27.3-42.6 **Gearbox:** Man/Auto **Boot:** 471-1405 **Seats**: 5 **Doors:** 4/5

Saab 9-3 1998 - 2002

Apparently the Swedish hatchbacked 3-Series according to the nomenclature. Bemusing engine and model combinations. HOT engine was just that but the 2.0 turbo petrol and diesels are the ones to own. Specifications were comprehensive so always worth aiming for an SE. Diesel fuel and oil leaks, worn timing chains, coil packs don't last long.

Engines: 2.0 • 2.2D **CO2:** 164-216g/km **MPG**: 31.4-45.6 **Gearbox:** Man/Auto **Boot:** 494/1314 **Seats**: 5 **Doors:** 5

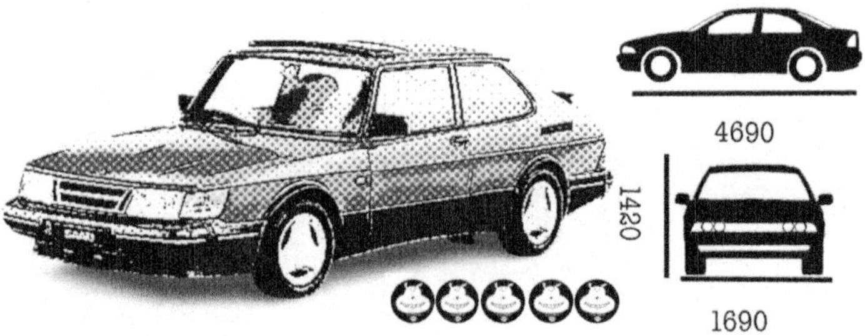

Saab 900 1984 - 1993

Ancient underpinnings, but a rock solid Banger legend that still made sense. Light Pressure Turbos in 1990, limited run models to '93 SE and Turbos. 16 valvers are the most fun. Six figure mileages are a given with these. Non Turbo may be underwhelming, but less Banger bother. Rattling timing chains, noisy camshaft, blowing turbos. Brilliant.

Engines: 2.0 • 2.0 Turbo **MPG**: 27.9-28.8 **Gearbox:** Man/Auto **Boot:** 650/1500 **Seats**: 5 **Doors:** 4/5

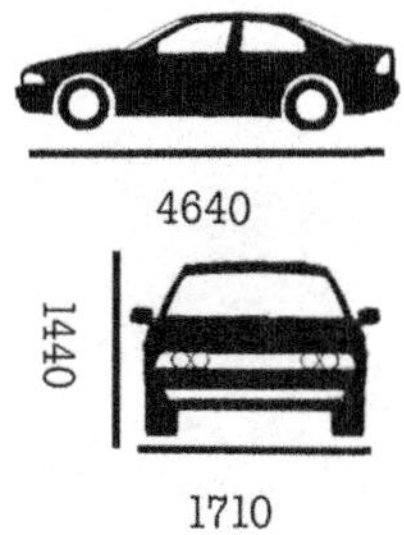

Saab 900 1993 - 1998

Essentially a Vauxhall Cavalier with a sheen of Scandinavian cool. Average to drive but it delivers in terms of specification and reasonably solid build quality, though not quite in old 900 terms. Massively complicated range of specs despite fairly short production run. Tired engines and gearboxes, quite tatty and unloved, not a legend.

Engines: 2.0•2.3•2.5 **MPG**: 29.9-32.8 **Gearbox:** Man/Auto
Boot:494/1413 **Doors**:3/5 **Seats**: 5

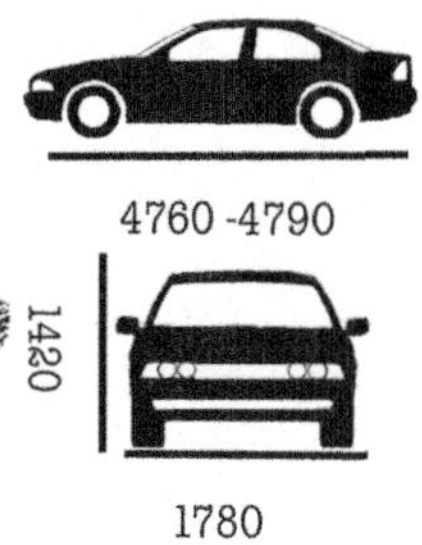

Saab 9000 1991 - 1997

The successful part of the ItalSwedo co-production. Excellent big saloon, or ginormous hatch. Got a larger body shell for this generation with uprated suspension. Lots of engine options and the Light Pressure ones are the best, V6 smoother and thirsty. Brilliant family Banger. Coil pack failure, air con issues, automatic 'box + suspension bushes.

Engines: 2.0•2.3•3.0 **MPG**: 29.2-32.3 **Gearbox:** Man/Auto
Boot:486-556/883 **Doors**: 4/5**Seats**: 5

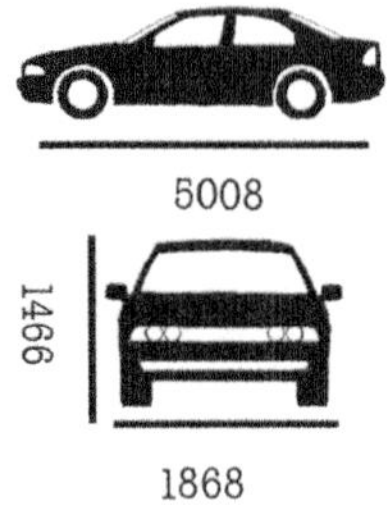

Saab 9-5 1997 - 2010

This will be the Swedish booted 5-Series according to the nomenclature. Huge inside and very comfortable, just not special. 2.0 and 2.3 and top of the range V6. 2001 revisions all round with diesel engines. 2005 facelift amounted to a headlamps with goth level mascara, still lovable. Coil pack failure, air con issues, automatic 'box + suspension bushes.

Engines: 2.0 • 2.3 • 1.9D • 2.0B • 2.3B **CO2:** 184-212g/km **MPG**: 31.7-41.5. **Gearbox:** Man/Auto **Boot:** 500 **Seats**: 5

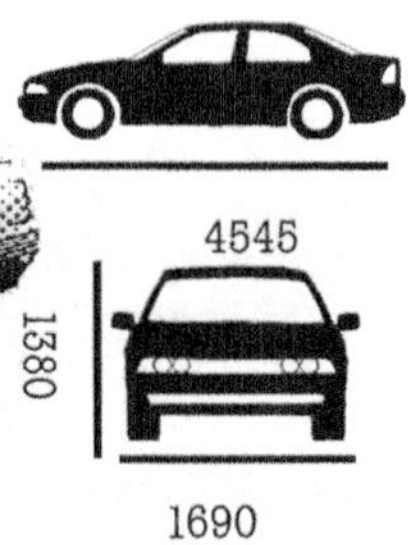

Subaru Legacy 1988 - 1994

Four wheel drive saloons at their rally winning best. Incredible performance on road and slightly off. Basic spec Ecos are spec free, 4-Cam is the mad one. Not very exec really and most rural buyers went for the estate anyway. Everything costs a fortune if broken. It will be wear and tear on the running gear if you can find one.

Engines: 1.8 • 2.0 • 2.2 **MPG**: 27.8-29.3 **Gearbox:** Man/Auto **Weight:** 1175 - 1335 **Boot:** 430/565 **Seats**: 5 **Doors:** 4

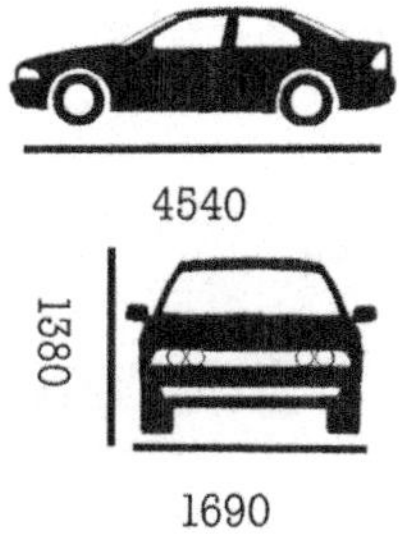

Subaru Legacy 1994 - 1999

Deathly dull styling compensated for by the handling, that flat engine noise and its overall reliability which comes at a price. LX Spec quite poor, run out Classic had air con, Lux Packs cheered thing up. As with all Scoobarus everything costs a fortune when busted. Most worn out now.

Engines: 2.2 • 3.0 **MPG**: 24.8-32.4 **Gearbox:** Man/Auto **Boot:** 405 **Weight**: 1155 **Seats**: 5 **Doors:** 4

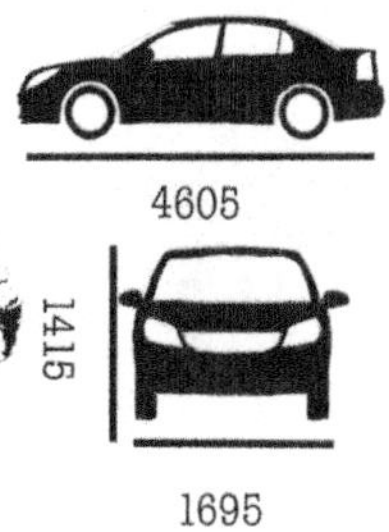

Subaru Legacy 1999 - 2003

Legendary four wheel drive saloon, but you don't have to get the mad one, the 2.0 litre engine GL is fine, but the 2.5 uses a little more fuel. Uglification revamp in 2001. Fixing them is expensive. Problems with the four wheel drive system only if neglected. Everything costs a fortune if broken. It will be wear and tear on the running gear.

Engines: 2.0 • 2.5 **CO2:** 215-228g/km **MPG**: 30.4-31.4 **Gearbox:** Man/Auto **Boot:** 433 **Seats**: 5 **Doors:** 4

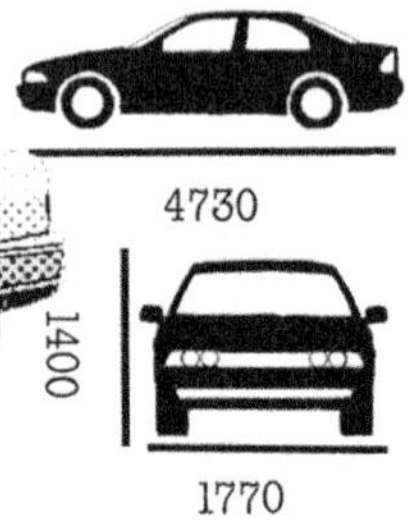

Toyota Camry 1992 - 1996

Big in the States, bought by pragmatists who wanted loads of equipment and utter reliability. Bags of room, and the 3.0 V6 GX has leather, cruise and headlamp washers. ABS and airbag from '93 so a pretty safe package. Very rare, the ones you find have wear and tear, otherwise it is brakes, tyres and suspension parts. Always expensive to fix.

Engines: 2.2 • 3.0 **MPG**: 24.8-32.4 **Gearbox:** Auto/Man **Boot:** 517 **Weight**: 1312 **Seats**: 5 **Doors:** 4

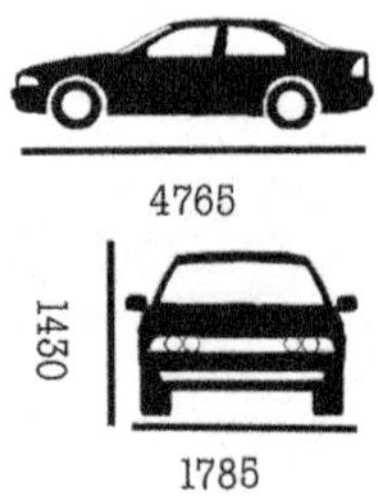

Toyota Camry 1996 - 2001

A simple enough line up with just a four and V6 cylinder engine in Standard and Sport flavours. 1999 update meant new lights and bumpers, but nothing else remotely interesting. The Camry is and was all about not breaking down. Wear and tear are the only issues and that will be brakes, tyres and suspension parts. Expensive to fix.

Engines: 2.2 • 3.0 **CO2:** 214-276g/km **MPG**: 24.4-32.1 **Gearbox:** Auto **Boot:** 524 **Seats**: 5 **Doors:** 4

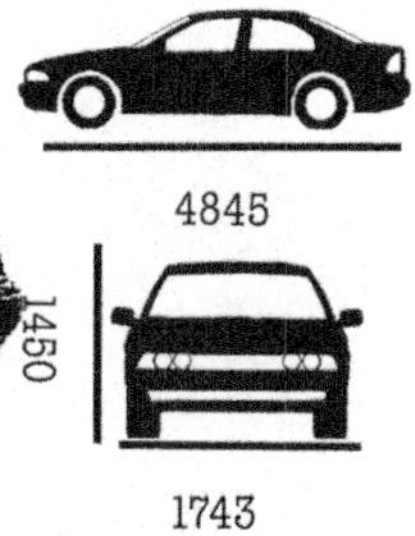

Vauxhall Carlton 1986 - 1993

A big family car, by 1990 ABS on the petrols, the Ls and GLs don't have much better get a CD and CDX. Diplomat and Diamond packs deliver upgraded trim of course. No Turbo diesel a slug and GSi the people's Lotus. Carlton. Not much loved at the time, so survivors will be worn out, suspension, brakes, engine and electrics.

Engines: 1.8 • 2.0 • 2.6 • 3.0 • 2.3D **MPG**: 24.4-41.6
Gearbox: Man/Auto **Boot:** 530/930 **Seats**: 5 **Doors:** 4

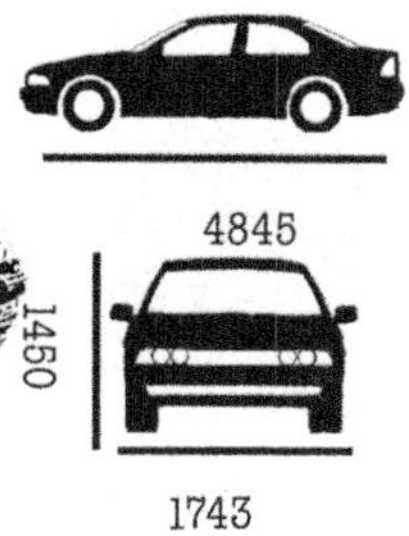

Vauxhall Senator 1987 - 1994

Upmarket Carlton powered by sixes and topped off with the egg box grille, entering the '90 catalytic converters. Certainly smooth and lots of kit the CD had electronic ride control, alloys and cruise. 24 valve 3.0i was also the ultimate cop car at the time. Worn suspension bushes, oil leaks, fuel injector seals, electricals with age.

Engines: 2.6 • 3.0 **MPG**: 25.4-32.7 **Gearbox:** Man/Auto
Boot: 530/930 **Seats**: 5 **Doors:** 4

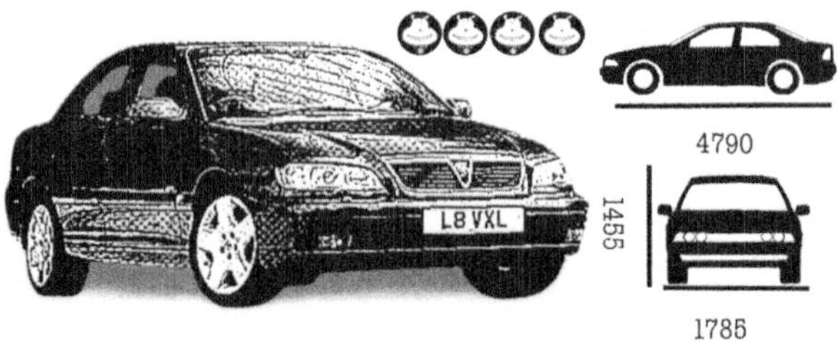

Vauxhall Omega 1994 - 2003

Made in Germany. Better looking from 1999 and the more powerful DTi diesels are the ones to have, unless you can run a V6 petrol for fun. A motorway missile with plenty of equipment, especially top end Elite and CDX. Worn suspension bushes, oil leaks, fuel injector seals, air conditioning and heating system.

Engines: 2.0 • 2.2 • 2.6 • 3.2 • 2.2D **CO2:** 192-284g/km
MPG: 23.9-39.8 **Gearbox:** Man/Auto **Boot:** 540

Volvo 240 1989 - 1991

It came from 1974 and by 1990 had the fuel injected engine and whilst the estate soldiered on for a few more years, the saloon was available in GL and GLT. From 1988 on they had galvanised bodies which has certainly helped but all the cross members and panels still need to be checked. So rotten bodies and engines that are smokey, cammy and the suspension goes bang. Saloons properly forgotten now.

Engines: 2.0 • 2.3 **MPG**: 27-29 **Gearbox:** Man/Auto **Boot:**395 **Weight:**1274 **Seats**: 5 **Doors:** 4

Volvo 850 1992 - 1996

Few wanted the notchback one, but that was there loss, here is a premium front wheel drive saloon that was great to drive. A bunch of great 5 cylinder engines too, from adequate to outrageous and frugal. Loads of spec level combinations which were always being revise. Tired 10V 2.0. smokey diesels and worn suspension. Faithful old thing.

Engines: 2.0 • 2.3 • 2.5 • 2.5D **MPG**: 25.2-46.7 **Gearbox:** Man/Auto **Weight:** 1473 **Boot:** 416 **Seats**: 5 **Doors:** 4

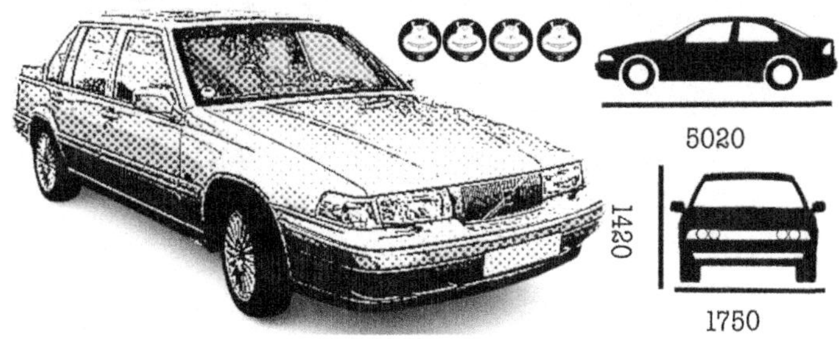

Volvo 940/960 1990 - 1998

Here is the 740/760 reinvented for the '90s and as ever the saloon is the least loved. Just a big collection of rectangles, until the suspension was tweaked in '92, then again in '94 with more chassis work and softening the styling rectangles. on wheels. Mind boggling spec/engine combos. Driven into the tarmac so worn suspension, smoky turbos and all that.

Engines: 2.0 • 2.2 • 2.3 • 2.5 • 3.0 • 2.4D **MPG**: 24.8-36.7 **Gearbox:** Man/Auto **Boot:** 491 **Seats**: 5 **Doors:** 4

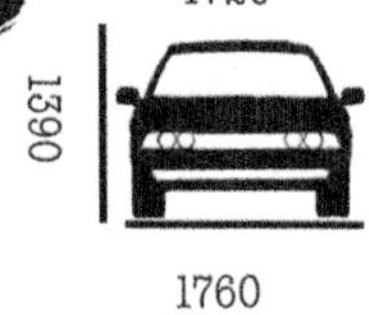

Volvo S70 1996 - 1999

The 850 was good and this is the Mark 2 version as Volvos got cuddlier. Saloon still the dull one but lots of engine options and madder R and T5 plus loads of option packs including winter, Comfort, SE and XT. Confusing, but at least whatever combo you end up with it is at least solidly reliable. The usual high mile issues that means rattles and smoke.

Engines: 2.0 • 2.4 • 2.5 • 2.5D **MPG**: 24.1-44.8 **Gearbox:** Man/ Auto **Boot:** 427-941 **Weight**: 1486 **Seats**: 5 **Doors:** 4

Bangersportscoupestargasmobile

The simple truth is that a performance car can very easily be Banger. It does the same job as a 78bhp 1.2 hatchback, but potentially it is a bit quicker. These are not hot hatches or scorching saloons, but models specifically designed to be just a little be special/different/usually pretty.

An Audi TT is a very good example of this. Underneath, like so many sports cars, is a rather 'ordinary car', a Golf. That is jolly good news because it means that long term it will be reliable and potentially cheaper to fix. I may have mentioned elsewhere just how these models rack up exceptional high mileages.

The exceptions are always going to be fabulously complicated creations like the Mitsubishi 3000GT, or a BMW 8 Series and any XK Jaguar. These are worth including because the used prices asked, certainly a decade or more ago were sometimes temptingly low. You might even get lucky with a pristine Mazda RX7 rotor surrounded by some moody paintwork that will polish up perfectly.

Mostly though, Banger Coupes are going to be dross aren't they? Ford Probes and Proton Coupes? Well actually the '90s was something of a golden age for quite brilliant two door cars. Honda had a whole brace of models from CRX and Civics and Accords to Legends which offered something for absolutely everyone. Pretty much the same with Nissan which had a whole bunch of sometimes quite odd NX, SX and ZX creations. The world clearly wanted to buy interesting, less practical sportsters.

Then if you want to consider properly weird ones there are those Scoobarus, the ones everyone forgot existed, because only half a dozen were sold. The SVX was a glimpse into the future and the XT is just charmingly different. Yes, dare to be different and get yourself one of those..... Bangersportscoupetargamobile.

Alfa Romeo Alfasud Sprint 1976 - 1990

What a beauty that just touched the beginning of the '90s. Ital Design at its best although rather more cluttered and black plastic adorned than the originals, but it remained a dream drive. Light, simple and flat four engined fun. They fell into disrepute as fast as any other Alfa and succumbed to rust and neglect as the pasta based wiring disintegrated..

Engines: 1.3 • 1.4 • 1.5 • 1.7 **MPG**: 34 **Top Speed:** 115
0-60: 9.1 **Boot:** 325 **Weight**: 915 **Seats**: 4 **Built**: 96,500

Alfa Romeo GTV 1996 - 2004

Head turning body by Pininfarina helped by a twin spark 2.0 litre under the bonnet from 1996, followed by a characterful 3.0 V6 in 1998. Standard specification was twin airbags and an alarm. Lusso pack though added air, leather and 16-inch alloys. Either looked after or falling to bits. Proper risky Alfa buy. MOT failure level rust, smokey engine.

Engines: 2.0 • 3.0 **MPG**: 24.1-30.7 **Top Speed:** 137-154
0-60: 8.9-9.6 **Gearbox:** Man/Auto **Boot:** 155 **Seats**: 4

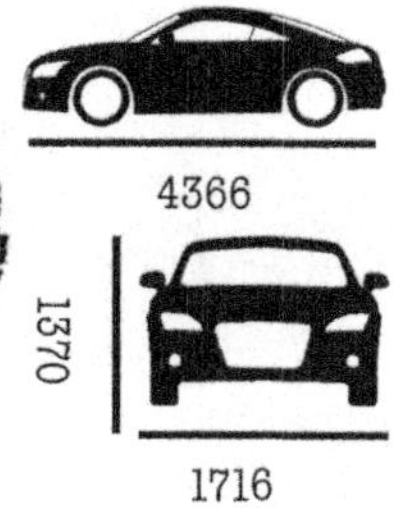

Audi 80 Coupe 1989 - 1996

Podgy but spacious, this is a full four seater, and the late '80s originals did not handle very assuredly until a '91 revamp which tickled the suspension. Top end was the 20V Quattro and S2, but the simpler ones were solid buys. Camshaft wear and knackered steering racks, mostly any seriously shabby ones will always cost a fortune to sort out.
Engines: 2.0 • 2.3 • 2.6 • 2.8 **MPG**: 27.4-32.9 **Top Speed:** 121-154 **0-60:** 5.9-12.9 **Gearbox:** Man/Auto **Boot:** 494

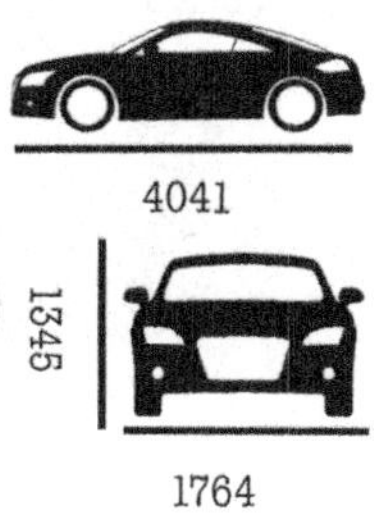

Audi TT 1999 - 2006

Automotive art. Thanks Audi for making traffic jams more stylish. Classic in waiting. However gushing you want to be this is just a Golf, but that is also the secret of its used car success. The choice is what state of tune for your 1.8T and whether or not you want to Quattro or not. Coil packs, ECUs, brakes, body damage. Can do seriously big mileages.
Engines: 1.8 **MPG**: 26.6-35.0 **Top Speed:** 141-155 **0-60:** 6.5-7.8 **Gearbox:** Man/Auto **Boot:** 220/490 **Seats**: 4

BMW 3 Series Coupe 1992 - 1998

No longer a two door saloon, it's a proper coupe' not quite as 1980's solid but probably slicker and certainly fine to own and drive. Usual slant of good fours, great sixes. Revamped in '96 a bit, Sports, SE and Lux packs. Pretty reliable, but fluid leaks, iffy trim is fragile, some electrics go 'bong'.
Engines: 1.6 • 1.8 • 2.0 • 2.5 • 2.8 **MPG**: 29.1-35 **Gearbox:** Man/Auto **MPH**: 121-147 **0-60**:12.7-7.3 **Boot:** 435 **Seats**: 4

BMW 3 Ci Series 1999- 2006

It needs to be repeated, repeatedly, these were the very last of the remotely handsome BMWs and the coupe is even better. Buy because of the looks, but also the sheer value for money of the bigger engined ones in particular. Motorsports will never Bangers of course, but there are some lovely 6 cylinder alternatives for less. Bore wear, suspension bushes.
Engines: 2.0 • 2.2 • 2.5 • 3.0 • 2.0D • 2.9D **MPG**: 28.5-31.7 **Top Speed:** 140-155 **0-60:** 6.5-8.2 **Gearbox:** Man/Auto
Boot: 410 **Seats**: 5

BMW 8 Series 1991 - 1999

Incredibly the slow selling continent crusher spent a while in used car wilderness waiting for brave or foolhardy buyers. Great V12 and V8 engines. Rather a lot of ECUs to go AWOL. Weighing in at a couple of tons means the suspension takes a bashing. Minors like water pumps and cam cover gaskets are the least of any owners worries. A world pf pain.

Engines: 4.0 • 5.0 • 5.2 **MPG**: 13-22 **Top Speed:** 155-170
0-60: 5.6 7.2 **Gearbox:** Man/Auto **Boot:** 320 **Built:** 30,600

Chevrolet Camaro 1998 - 2004

Not something British coupe buyers ever asked for, which might explain why a handful were sold. Worth bearing in mind because a scrappy one might tempt you. Optional pop out targa panels mean that you can appreciate the 5.7 V8 soundtrack rather better. Otherwise pushed along by a more workaday V6. Parts are cheap, freight may not be. Unstressed, might have seized brakes and minor rust.

Engines: 3.8 • 5.7 **MPG**: 20.9-22.8 **Top Speed:** 125-157
0-60: 5.5-8.5 **Gearbox:** Auto **Boot:** 187 **Seats**: 2+2

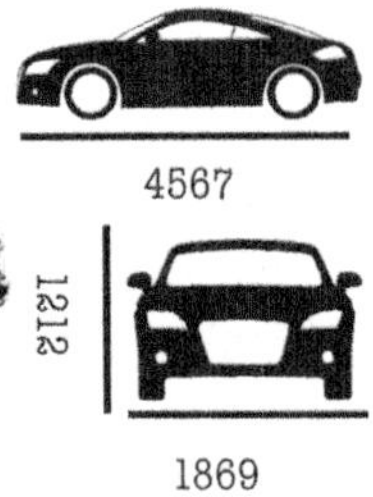

Chevrolet Corvette 1998 - 2002

The C5. Unlikely ever to be a Banger, but worth a mention, just in case there is a damaged repairable or partly immolated one to consider. Targa roof was an option, right hand drive less so. Left hook only, otherwise most things are standard. Sorting a scruffy one out may not cost a fortune in cheap mechanical parts. We'll forget about the C6.

Engines: 5.7 **CO2:** 299g/km **MPG**: 21.6 **Top Speed:** 170
0-60: 5.1 **Gearbox:** Auto **Boot:** 187 **Seats**: 2

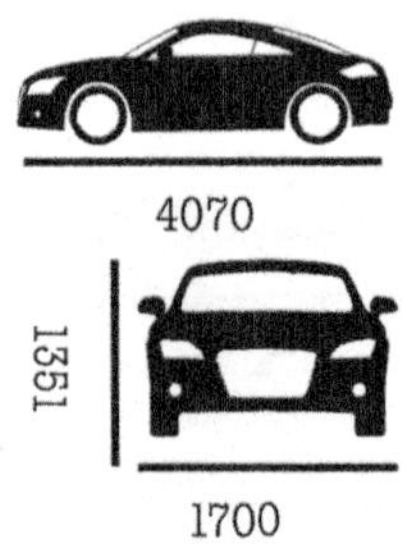

Citroen ZX Coupe 1992 - 1997

It's a more up to date Pug 306, but not half as stylish or desirable. It's a hatch, but they call it coupe so its here. Consequently fairly spacious and practical, so you can do the shopping with the family. The Volcane is the exciting one. Still fixable. ECU issues and electricals along with all other '90s Citroens. Not hip and happening, so most have died.

Engines: 1.4 • 1.6 • 1.8 • 2.0 • 1.9D **MPG**: 33.6-47.4
Top Speed: 107-137 **0-60:** 11-8 **Boot:** 324/926

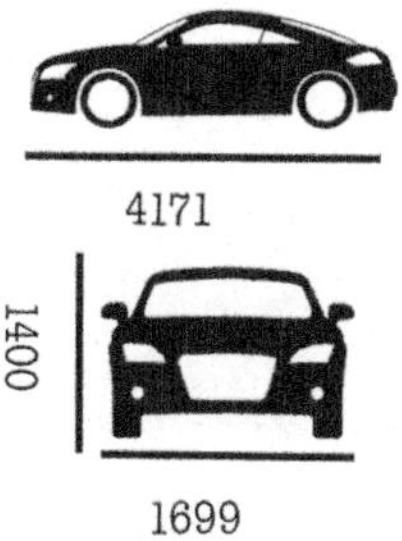

Citroen Xsara Coupe 1998 - 2004

Dull but worthy range of vehicles that will last a lifetime. Just a booted two door. The HDi 110 does have a bit of attitude which delivers economical entertainment. There is a traditional hot hatch in the shape of the 167bhp VTS. No one really wants them. But they should. Still fixable. ECU issues, some late '90s diesel engine . Diesel clutches weak.
Engines: 1.6•2.0•2.0D **MPG**: 30.4-54.3 **Top Speed:** 121-137 **0-60:** 30.4-54.3 **Gearbox:** Man/Auto **Boot:** 480/595

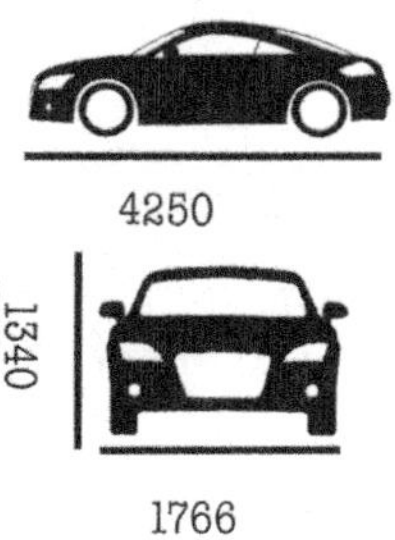

Fiat Coupe 1994 - 2000

Bang on '90s, wonderful at the time, inspired in retrospect, however, quality was not exactly wonderful. Great front wheel drive fun though. Avoid sitting in the back if you can, always insist on the driving seat. Basic 2.0 16V is the Banger, Turbo is potentially more troublesome. So smoke, rust, electricals, brakes, costs a lot to sort, but worth it.
Engines: 2.0 **MPG**: 28-30.7 **Top Speed:** 129-155
0-60: 6.5-9.2 **Weight**:1285 **Boot:** 295 **Seats**: 4

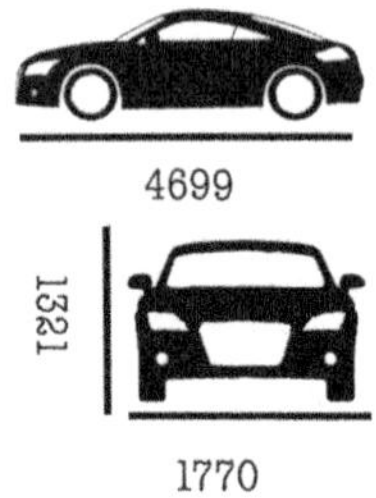

Ford Cougar 1998 - 2002

Those who wanted it to be the Capri were either retired or dead. Simply a collection of sharp shapes and ordinary engines. Unloved and effectively a less practical Mondeo. Good standard spec though for safety and comfort, whilst a V6 adds leather. Overlooked gem. Electrics and smokey engine, suspension and brake wear.

Engines: 2.0 • 2.5 **MPG**: 29.7-34.4 **Top Speed:** 130-140
0-62: 7.5-9.6 **Gearbox:** Man/Auto **Boot:** 351 **Seats**: 4

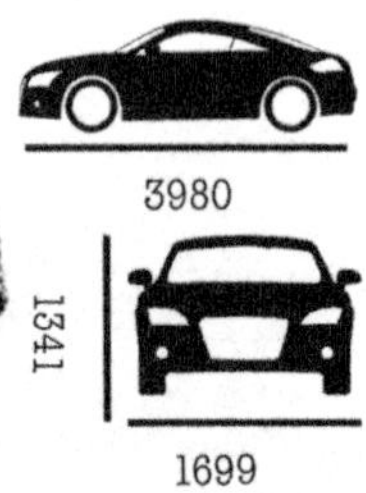

Ford Puma 1997 - 2002

The Fiesta was great to drive and popping a fat feline body on top was greatly appreciated. The 1.7 was powerful enough although the 1.4 rather underwhelming. There was a limited edition Racing but the 1.6 was better. Fun to drive, reliable and cheap to run, easy to get parts for, but sadly rampant rust is the real killer.

Engines: 1.4 • 1.6 • 1.7 **MPG**: 38.2-38.7 **Top Speed:** 117-122
0-60: 8.6-10.8 **Gearbox:** Man **Boot:** 241/725 **Seats**: 2+2

Ford Probe 1994 - 1997

Terrible name and a fairly terrible brand new car, overall just dull. Used though, prices dropped like a breeze block, plus very loaded specs, especially the V6. Uprated in mid '95 with new seats, alloys, then bigger spec at the end of the year with leather and air for the 2.0. Electricals, general neglect and now indifference damn it to Banger obscurity.

Engines: 2.0 • 2.5 **MPG**: 27.5-35.3 **Top Speed:** 127-136 **0-60:** 8.5-10.6 **Gearbox:** Man **Boot:** 337 **Seats**: 2+2

Honda Accord Coupe 1998 - 2001

Large comfy, anonymous and inoffensive two door barge. That's a shame as it is brilliantly engineered and totally reliable. 2.0 ES uses less petrol than the 3.0 V6. Both have got big boots for the owners and their golf clubs, but also luggage. Great for long distance, two up travel and long term ownership. Brakes and suspension give problems.

Engines: 2.0 • 3.0 **MPG**: 26.4-31.4 **Top Speed:** 129-139 **0-60:** 8.8-10.0 **Gearbox:** Man/Auto **Boot:**467 **Seats**: 5

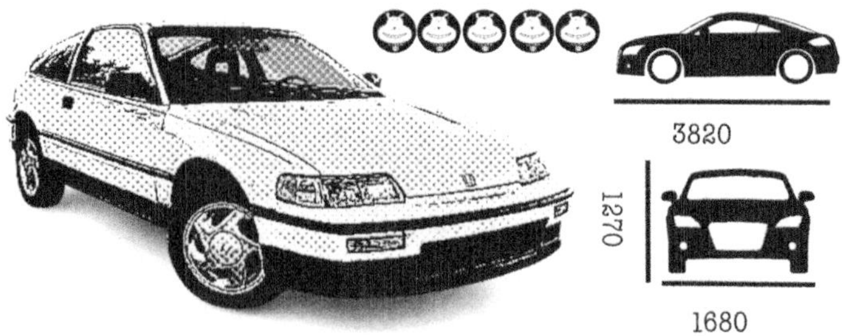

Honda Civic CRX 1988 - 1992

Stubby and rather brilliant road rocket. Arguably a hot hatch rather than coupe, but it is sporty enough justify a stand alone entry. VTEC engine is all very exciting at 7000rpm and the chassis is up to the job. Many were Max Powered to oblivion. Caution needed then, plastic interior survives even if rotting body doesn't. Wonderful.

Engines: 1.6 **MPG**: 42.2- **Top Speed:** 137 **0-60:** 7.3
Gearbox: Man/Auto **Boot:** 197 **Seats**: 2 + 2

Honda Civic CRX 1992 -1997

Not the purposeful rocket it used to be (but there is the VTi), complete with the complicated but optional super clever electric roof from '95. Pop out targa for the hardcore cheapskates. Drivers airbag in '95 and ABS on the top models. Revised suspension in '96. Disappointing but only compared to the original CRX, it's reliably Honda.

Engines: 1.6 **MPG**: 36.6- **Top Speed:** 118-131 **0-60:** 9.3-7.8
Gearbox: Man/Auto **Boot:** 300 **Seats**: 2

Honda Civic Coupe 1994 - 2003

Not much to look at, but the engines are universally excellent and obviously unlikely to break down. Reasonable amounts of room inside and at least four can get on board without long term physical damage. VTEC engines for the enthusiast. Ideally, buy from someone who does not obviously identify as a Fast and Furious fan.

Engines: 1.7 **CO_2:** 159g/km **MPG**: 42.2- **Top Speed:** 125 **0-62:** 9.6 **Gearbox:** Man/Auto **Boot:** 325 **Seats**: 5

Honda Prelude 1987 - 1992

Third generation wedge, sort of micro NSX which was a few years away, but had mechanical four wheel steering tech as an option. Pop up lights always delightful. Relatively light and consequently handles sweetly with those wonderful VTEC engines. Few took it that seriously or acted fast enough to stop the rot. Perfect, cool late '90s Banger.

Engines: 1.8•2.0 **MPG**: 29.4 **Top Speed:** 118 **0-60:** 9.2 **Gearbox:** Man/Auto **Boot:** 379 **Seats**: 2+2

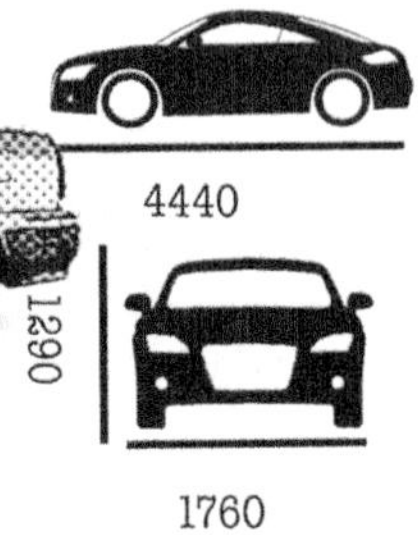

Honda Prelude 1992 - 1996

Away with the pop up headlamps, but in comes a more conventionally shaped coupe. 4WS is still a hardware thing, good equipment, cruise, alloys, electric mirrors. 1994 bigger boot space saver spare and revamped dashboard and some new alloys. The 2.2 VTEC is the enthusiast choice, but Banger coupe lovers want standard Honda engine brilliance.
Engines: 2.0 • 2.2 • 2.3 **MPG**: 31.4-32.3 **Top Speed:** 126-140 **0-60:** 7.1-9.2 **Gearbox:** Man/Auto **Boot:** 278 **Seats**: 2+2

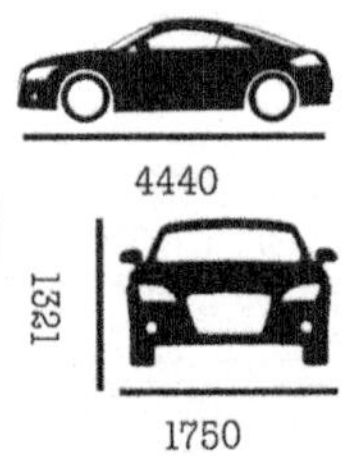

Honda Prelude 1997 - 2001

Buyers know that it will be well built and never break down. The Prelude is reasonably interesting with four wheel steering and those VTEC engines. As ever buy from an owner that cared, or at least had it over serviced, whatever the mileage. Getting rare. Just wear and tear, ideally not a noisy and rough VTEC.
Engines: 2.0 • 2.2 **MPG**: 29.5-30.7 **Top Speed:** 126-140 **0-60:** 7.1-9.2 **Gearbox:** Man/Auto **Boot:** 284 **Seats**: 2+2

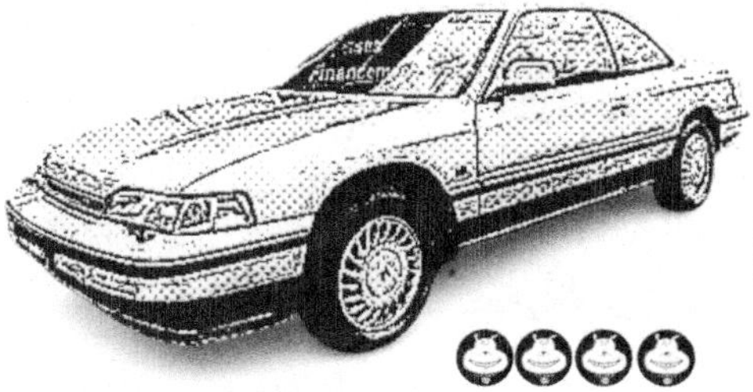

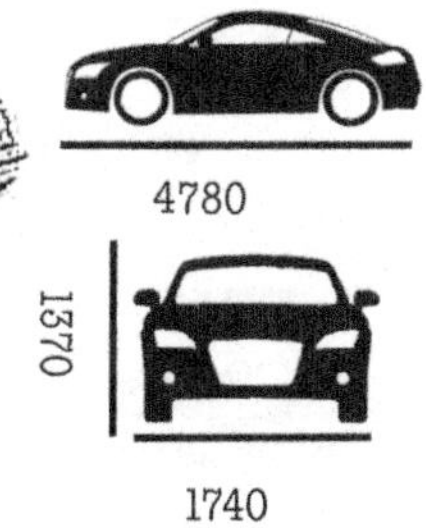

Honda Legend Coupe 1988 - 1991

It's an Accord+ and Rover developed in Coventry with Rover 800 inner body panels, but a 2.7 Honda V6 under the bonnet. Comfortable and well equipped, most had leather and electric everything, but just did not connect with the buying public as a big luxury Honda with just two doors. Honda elements fine, everything else rusts to bits.

Engines: 2.7 **MPG**: 28 **Top Speed:** 132 **0-62:** 8.1
Gearbox: Auto **Boot:** 440 **Seats**: 4

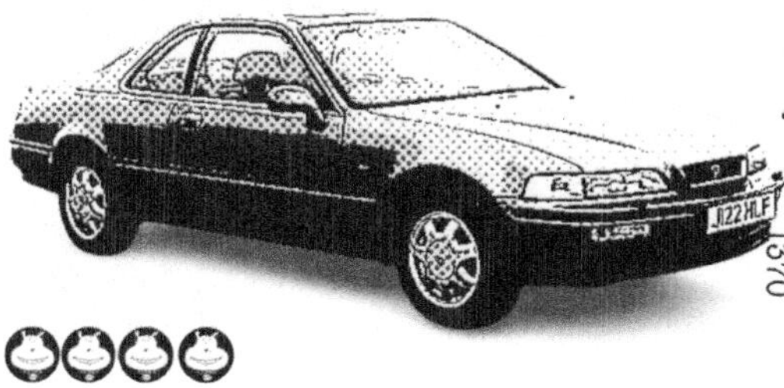

Honda Legend Coupe 1991 - 1995

Revamped Legend and buyers were just as indifferent despite the bigger V6. More interest Stateside where it was Acura branded. Built to be a large Lazy Boy sofa of thing and excelled at that. ABS, power steering, cruise, air and leather. Driver's airbag, passenger one from 1993. Rarely went that wrong, buy from retired company directors.

Engines: 3.2 **MPG**: 26 **Top Speed:** 140 **0-60:** 8
Gearbox: Man/Auto **Boot:** 427 **Seats**: 4

Hyundai Accent-Coupe 1994 - 1999

This is nothing more than a three door hatchback with an ability to pull the skin of a rice pudding. Just. Had a remote fuel flap, internally adjustable door mirrors and a rear wash/wipe. In '94 it got a high mounted stop lamp and a radio cassette. That was as exciting as it was ever going to get. Being a Hyundai, nothing to worry about.

Engines: 1.3 **MPG**: 46 **Top Speed:** 109 **0-60:** 11.7 **Gearbox:** Man/Auto **Boot:** 342 **Seats**: 4

Hyundai S-Coupe 1990 - 1996

An unremarkable thing, the Lsi had alloys and bucket front seats whereas the GSi glued on a spoiler and some bits of trim. Second generation from '92 had revised suspension, electric mirrors and now there was a turbo. Turbo SE in '95 had lower suspension. Bought on the basis of a low price they had blowing engines/scrappy bodies and cost buttons.

Engines: 1.5 • 1.5 Turbo **MPG**: 32.1-35.4 **Top Speed:** 107-121 **0-60:** 9.1-12.5 **Gearbox:** Man/Auto **Boot:** 261 **Seats**: 4

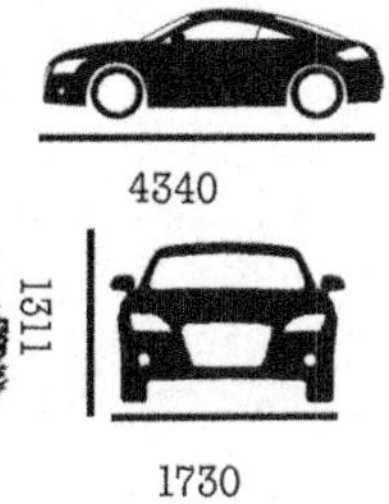

Hyundai Coupe 1996 - 2002

Hasn't got the badge or the looks, but it will at least fail to break down then be easy and cheap to look after. So not much going for this on the face of it, but as a set of stop gap wheels it will be better than alternatives. There is only going to be rust otherwise wear and tear to brakes, exhaust, tyres and other consumables.

Engines: 1.6 • 2.0 **MPG**: 31.7-36.7 **Top Speed:** 120-125
0-60: 8.6-11.2 **Gearbox:** Man/Auto **Boot:** 362 **Seats**: 2+2

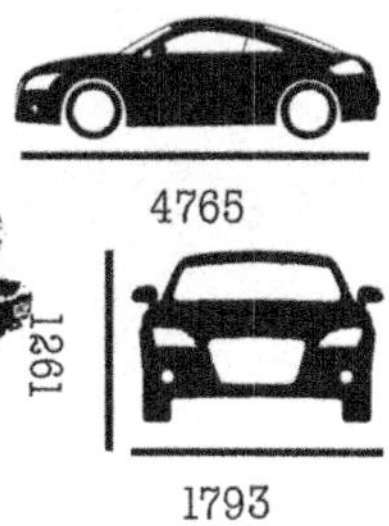

Jaguar XJS 1991 - 1996

Last gasp for the '70s survivor and these revised '91s have galvanised bodies so not to be confused with the old rotters. Despite simplified construction they still oxidise. Revised AJ6 engine is better and became the AJ16 in '94. Low miles can be fatal, they need exercise, so smoke, coolant, oil leaks, worn suspension and radius arms. Were cheap for a while...

Engines: 4.0 • 5.3 • 6.0 **MPG**: 18.8-24.3 **Top Speed:** 141-161
0-60: 6.6-7.8 **Gearbox:** Auto **Boot:** 480 **Seats**: 2+2

Jaguar XK8 1996 - 2005

Quite a lot of ancient XJ-S in its construction and various parts can cost a fortune. So Banger money, for what is a classic, but hardly something to smoke around in without fix it skills or a thick wallet. Biggest later update was in 2002 with bigger engines, Xenon headlamps and fresh alloys, then a final facelift in 2004. Electrics, ECUs, rust, suspension.

Engines: 4.0 • 4.2 **MPG**: 22.2-24.1 **Top Speed:** 155 **0-60:** 4.4-7.1 **Gearbox:** Auto **Boot:** 326 **Seats**: 2+2

Mazda 626 Coupe 1987 - 1992

A handsome and purposeful looking thing that had four wheel steering in some markets, in the UK it was a GTi with a front and rear spoiler. ABS and sports seats and rebadged as a plain GT in 1990. The rarest of the rare. Usual Mazda levels of unbreakdownability. Certainly rusts and faces absolute extinction due to a lack of spares.

Engines: 2.0 **MPG**: 32 **Top Speed:** 129 **0-60:** 9.4
Gearbox: Man/Auto **Weight**: 1175 **Boot:** 457 **Seats**: 4

Mazda MX-3 1991 - 1998

'90s oddity, a micro coupe available with a titchy V6 which is ace. Auto only 1.6 with the manual V6, had alloys, electric mirrors and colour coding. Side impact beams in '93 then in '94 new seats and trim, big news though was manual 1.6 in '96. Underrated, reliable and the perfect alt-Tigra/MR2. Best pocket sized Banger of the era, just too hairdressery.

Engines: 1.6 • 1.8 **MPG**: 30.1-38.1 **Top Speed:** 105-126 **0-60:** 8.5 - 13.8 **Gearbox:** Man/Auto **Boot:** 289-390**Seats**: 4

Mazda MX-6 1991 - 1998

Mazda's very own Probe, but better to look at, with more kit and same smooth V6. Came with alloys, electric mirrors and cruise. 1996 saw leather and air con as standard. Had just as much penetration in the market as the Probe. Just a Touring car rather than a sportster, but the nest bit its failure to throw a fit because it is a Mazda.

Engines: 2.5 **MPG**: 29 **Top Speed:** 137 **0-62:** 8.2 **Gearbox:** Man **Boot:** 404 **Seats**: 4

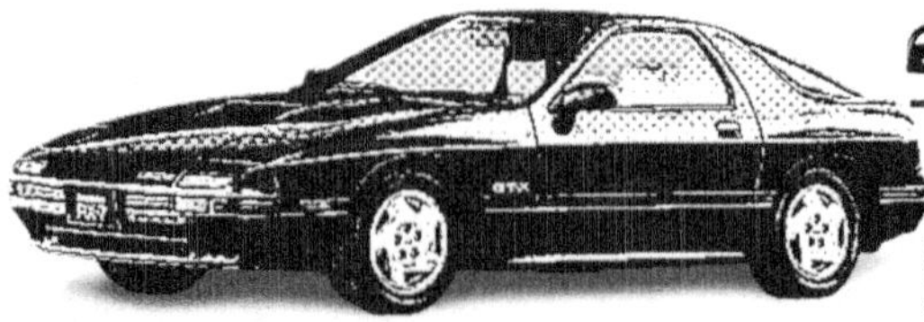

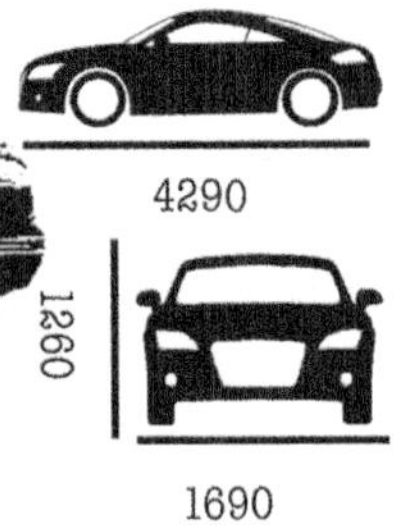

Mazda RX-7 1985 - 1992

Buyers are inevitably attracted to the oddest of coupes, but should of course run the other way. Despite being well built, the Rotary engine is a harbinger of oil leaks and expensive doom sometimes with an added turbo. Despite the cleverness of it all and the initial depreciation, it tempted some to sign up for Rotund Bangerdom.

Engines: 1.3 **MPG**: 24.6 **Top Speed:** 131 **0-60:** 8.4
Gearbox: Man **Boot:** 200 **Seats**: 4

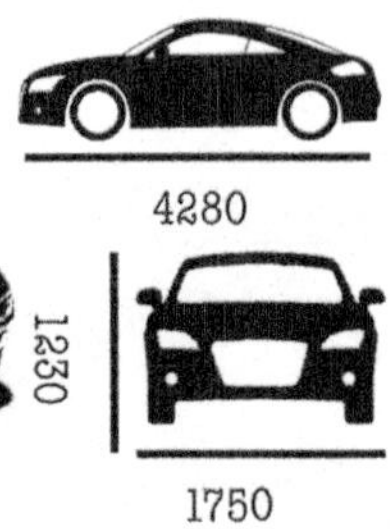

Mazda RX-7 1992 - 1996

Torn about whether to include this, but the depreciation for the best handling and most potent RX-7 was epic, which drew a lot of budget supercar hunters in, later by the grey import route. Compression testing is the only way to buy with some sort of foreknowledge of Rotary doom. Sometimes cheap, never really a bargain unless looked after proper.

Engines: 1.3 **MPG**: 25 **Top Speed:** 155 **0-60:** 5.3
Gearbox: Man **Boot:** 122 **Seats**: 2

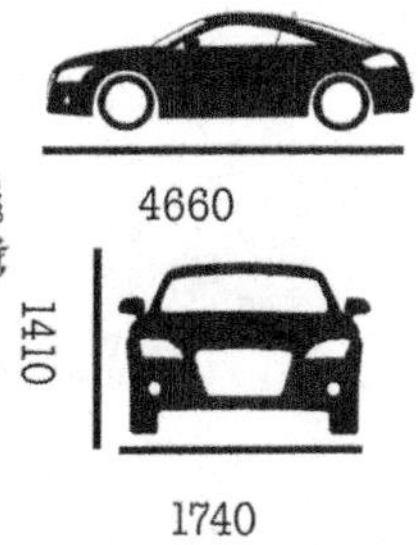

Mercedes CE / E 1988-1996

Accountant's idea of a Grand Tourer in W124 wrapping. From August '93 badged as Es with mild facelift and spec upgrade for 1995 which meant walnut trim, hole alloys and headlamp wash/wipe. Model run outs in '96 got cruise and anti-skid. Other worldly build quality, but not immune from early ECU failure and shonky auto boxes

Engines: 2.2 • 2.3 • 3.0 • 3.2 **MPG**: 25.3-33.3 **Top Speed:** 126-147 **0-60:** 7.5-10.6-**Gearbox:** Man/Auto **Boot:** 480 **Seats**: 4

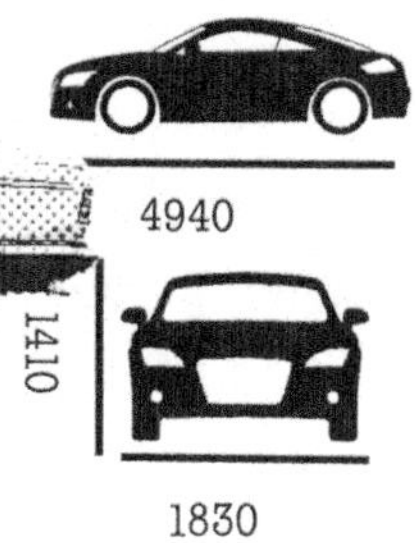

Mercedes SEC 1979 - 1992

Colossally accomplished and over the top and identified by nerds as C126. Indeed they would claim its a mildly reworked W116, so despite the high techness added over a couple of decades (ABS, cruise) it is relatively uncomplicated. Putting one right will make a large fortune quite small. Probably worth it though.

Engines: 2.8 • 3.0 • 5.0 **MPG**: 20 **Top Speed:** 149 **0-60:** 7.2 **Gearbox:** Auto **Boot:** 510 **Seats**: 4

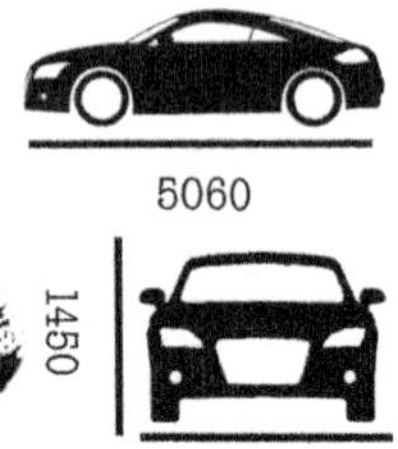

Mercedes SEC/S Coupe/CL - 1992 - 1998

More of the same, but in C140 wrapping and many claim the best S ever and quite rare in 2 door form, and renamed every few years. All had diamond cut alloy wheels, double glazing, 'close assist' doors (and boot lid), dual zone climate control and, on the 500 and 600, full Nappa leather and its the first new Merc with a radio as standard. Bundles to fix.
Engines: 4.2 • 5.0 • 6.0 **MPG**:20 - 24 **Top Speed:** 152- 155 **0-60:** 6.3-8.3 **Gearbox:** Auto **Boot:** 505 **Seats**: 4

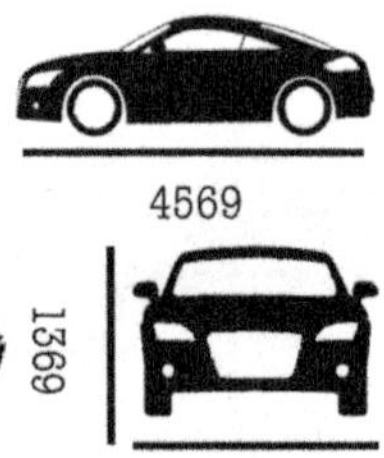

Mercedes CLK 1998 - 2003

Arguably the pick of Merc's huge range of stylish two doors, although the earlier models not built in the best era for quality control. So there may be rust and other issues. Massive range of engines from the ordinary 2.0 petrol to the V8s and CDI diesels. Beware rust and electrical issues.
Engines: 2.0 • 2.3 • 3.2 • 4.3 **MPG**: 24.2-28.8 **Top Speed:** 129-155 **0-60:** 5.4-9.7 **Gearbox:** Auto **Boot:** 419 **Seats**: 4

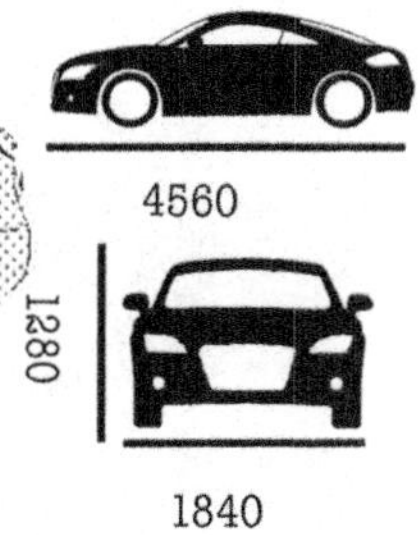

Mitsubishi 3000GT 1992 - 1996

Technological Tour de Force as they say. Far East supercar with four wheel drive, four wheel steering and a six speed manual 'box from '96. Apart from changing gear it does everything for the driver. It's a more complicated Supra and a less special NSX, which became something of a bargain and grey import darling for a while. It remains complicated.

Engines: 3.0 **MPG**:25 **BHP**:282 **Top Speed:** 155 **0-60:** 5.9
Gearbox: Man **Boot:** 120/554 **Weight:** 1730

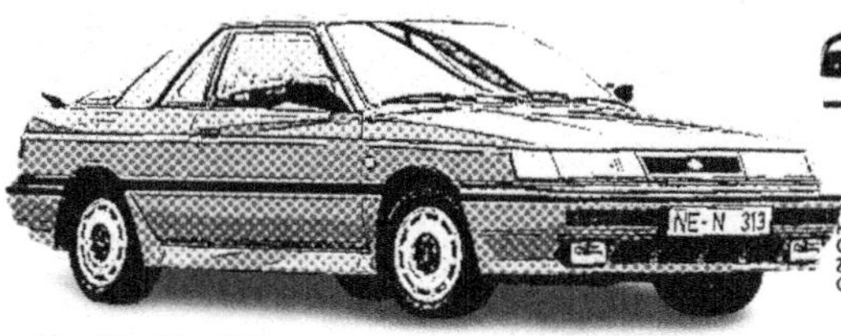

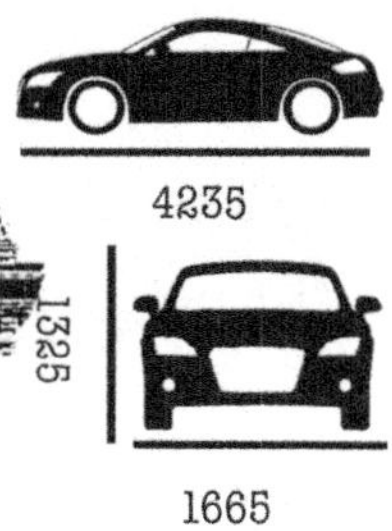

Nissan Sunny ZX Coupe 1987 - 1991

This is important, the ZX had a twin cam engine that was fairly lively, whereas the GSX was a single cam job that was a bit slower and took 3 seconds more to 60. 1.8 from 1989 bigger brakes, uprated springs and clutch plus alloys and bronze tinted glass. Sunroof deleted in 1990. An unremarkable and forgotten, it is just the small details now.

Engines: 1.8 **MPG**:34.7 **Top Speed:** 125 **0-60:** 8.5
Gearbox: Man **Boot:** 409 **Seats**: 4

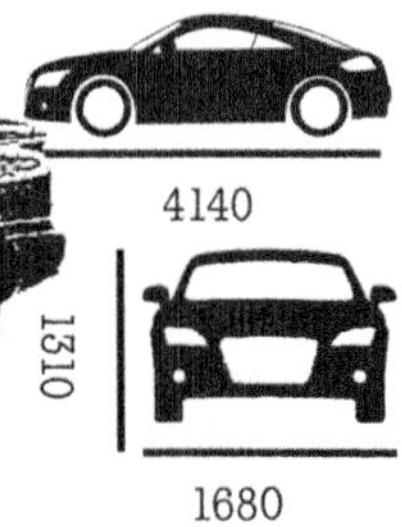

Nissan 100SX 1991 - 1995

A sort of sexy Nissan Sunny, available as a tin and targa top. Fugly looking thing but fun in a zero cred manner. Biggest revamp in 1993 with multi point engines and side impact beams. Nissan badge means it won't break down, but most neglected and abused. That targa roof will leak, most rotten and dented panels impossible to replace.

Engines: 1.6 **MPG**: 39.7-40.1 **Top Speed:** 115--118
0-60: 9.9-10.4 **Gearbox:** Man/Auto **Boot: 328 Seats**: 4

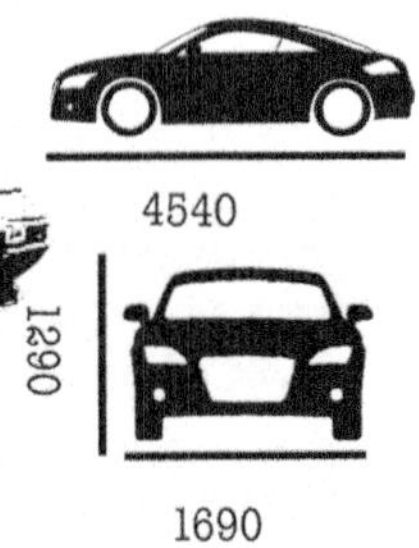

Nissan 200SX 1989 - 1994

Rear wheel drive, a turbo and pop up headlamps, all the required '90s tick boxes ticked there. Executive model had air and leather gear knob and steering wheel. Modded by Nissan over the years, ABS and LSD for '92, air con a year later with leather and a CD. Many never made it out of the decade without serious modification or damage.

Engines: 1.8 **MPG**: 30.3 **Top Speed:** 137 **0-60:** 6.7
Gearbox: Man/Auto **Boot:** 260 **Seats**: 4

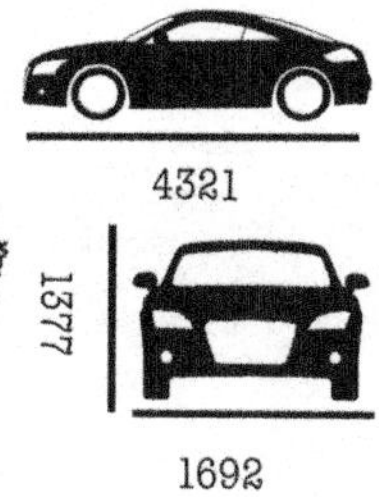

Nissan 200SX 1994 - 2001

Very old school Datsun from the early days of drifting. Originally bought by old fellas with chest hair then modified later by teenagers. High spec Tourings seem to make up most of the survivors. The secret is to dodge the Fast & Ferocious ones that have been chipped and over boosted to infinity and very much beyond. Rust, mods, crashes etc...

Engines: 2.0 **CO2:** 267g/km **MPG**: 21.6 **Top Speed:** 146 **0-60:** 7.1 **Gearbox:** Man/Auto **Boot:** 240/397 **Seats**: 4

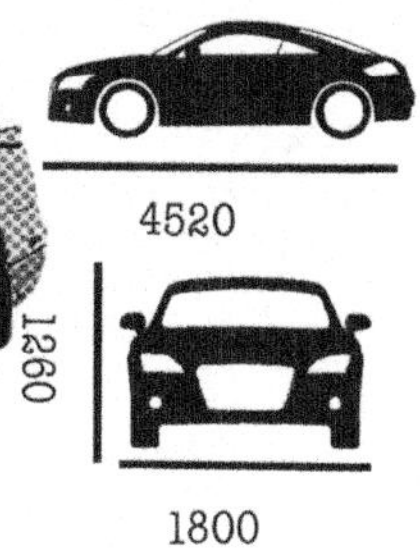

Nissan 300 ZX 1990 - 1994

Essentially a working man's supercar. Relatively affordable and worryingly complicated. Four wheel steering and a turbo. Came with a Targa in the UK and a ton of standard kit including air and cruise. Later on 'Luxury' spec meant leather and a CD. Those are details, the really important thing is what electrics and ECUs are still operational.

Engines: 3.0 **MPG**: 24.7 **Top Speed:** 155 **0-60:** 5.7 **Gearbox:** Man/Auto **Boot: 325 Seats**: 2

Peugeot 406 1997 - 2003

Rather elegant, but the Pug badge never helped matters, so depreciation made it even more attractive. 2.0 and 3.0 V6 petrols, but the excellent HDi diesels are the practical picks plus they all have split rear seats. Good specification levels even pre SE. Those drop down windows cost a bit to fix if playing up. Corroded radiators, rusty brake discs, engine ECU issues, suspension bushes.

Engines: 2.0 • 3.0 • 2.2D **MPG**: 25.9-44.1 **Top Speed:** 146
0-60: 7.9-10.9 **Gearbox:** Man/Auto **Boot:** 391 **Seats**: 4

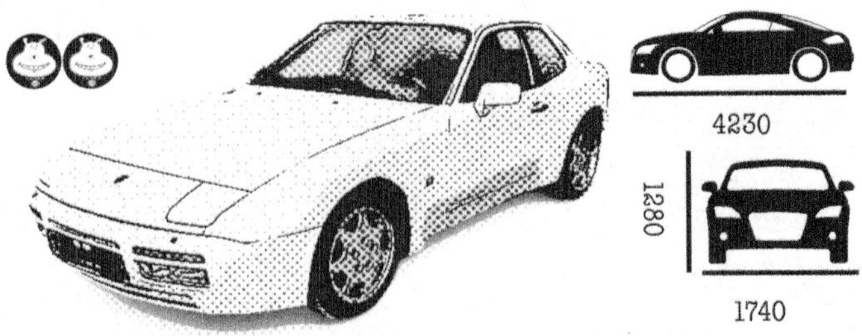

Porsche 944 S2 1988 - 1993

Odd to think that these were once Bangers, a fate the 968 evaded, but these routinely fell into the wrong hands despite having the engine in the right place. S2 the most numerous and the turbo unless exploded maintained desirability. There is plenty to look out for, they will rust, belts, chain and tensioner all need recent replacement. Don't relax.

Engines: 2.7 • 3.0 **MPG**: 28.9-29.2 **Top Speed:** 149-161
0-60: 5.6-6.7 **Gearbox:** Man **Boot:** 250 **Seats**: 4

Porsche 928 S41986 - 1992

Incredible that in the early '90s recession the values of these plummeted. Made sense to buy and mothball, but neglecteds will ruin your life. Production with the manual only GT went on until '95. Unitidy S4? The fact that the air con does not work would be the least of your worries. Electrics always fun to sort out. Despite galvanisation, will rust. Nightmare.
Engines: 4.5 **MPG**: 25.9-44.1 **Top Speed:** 168 **0-60:** 5.9-
Gearbox: Auto **Boot:** 200/400 **Seats**: 4

Proton Coupe 1997 - 2001

The badge, or styling does not help, so the real bonus here is back to basics Mitsubishi Lancer related engineering and zero desirability. It also has a boot spoiler and Recaro seats. Decent basic spec for the time and a split rear seat. Quite likely to last until the end of time. Reliable, maybe check for head gasket, oil leaks and just worn out suspension parts.
Engines: 1.8 **CO2:** 208g/km **MPG**: 32.7 **Top Speed:** 126
0-60: 9 **Gearbox:** Man **Boot:** 400 **Seats**: 4

Renault Megane Coupe 1996 - 2003

Popular compact sized coupe with a smallish boot and quite honestly only room for kiddies in the back. Properly sporty when driven hard it had a major restyle in 1999. Came with a wide range of petrol engines. Easy to sort models and no diesels keeps it simple. Gearbox and electrics. Clutch repair difficult, coil packs and flywheel sensors.

Engines: 1.4 • 1.6 • 2.0 **MPG**: 37.2-43.5 **Top Speed:** 116-132
0-60: 8.6-11.4 **Gearbox:** Man/Auto **Boot:** 289 **Seats**: 5

Rover 200 Coupe 1992- 1999

Tomcat by development nickname and can spend a lot of time at the vets (garage) being sorted. T-Bar roof and sensational turbo performance are the plus points. The Honda 1.6 is the best banger option, but can cost more than average to fettle. ABS failure, Rover 2.0 engine especially early M16, 1.8 K Series head gaskets. There's a lot to check.

Engines: 1.6 • 1.8 • 2.0 **MPG**: 32.3-38 **Top Speed:** 120-150
0-60: 9.5-6.2 **Gearbox:** Man/Auto **Boot:** 300-610 **Seats**: 4

Rover 800 Coupe 1992 - 2000

If the Honda Legend doesn't do it, pay more for the less well realised, fairly average to drive, Edwardian interior equipped 800. They will all rust the 2.0 likes to leak oil, the gearbox costs a bomb to fix and at the very least the hydraulic tappets on the V6 will get noisy. Expensive when new, then cheap, then expensive again, because nostalgia.

Engines: 2.0 • 2.5 **MPG**: 27.3-31.5 **Top Speed:** 125-143 **0-60:** 10-7.3 **Gearbox:** Man/Auto **Boot:** 470 **Seats**: 4

Subaru SVX 1992 - 1997

Spectacular oddity. It looked like the future and drove like it too with a flat six cylinder and all wheel drive. Not ideal Banger ingredients, but became the cheapest used sub supercar you could buy. Everything costs a fortune to sort and source. A tatty one better and possibly less troublesome to fix than an old fella's XJS. Coolest Coupe of all time.

Engines: 3.3 **MPG**: 27 **Top Speed:** 146 **0-60:** 8.6 **Gearbox:** Auto **Boot:** 240/466 **Seats**: 4

Subaru XT Coupe 1985 - 1990

It's the second oddest Subaru coupe, is a sort of SVX light, but looks like an '80s vision of the near future. Interestingly it can be switched to 4WD or simply engaged via kick-down. Permanent four wheel drive from '87 when automatic became and option. Made it to the '90s and the Bangerpedia wanted to celebrate that. Buying advice as SVX.

Engines: 1.8 **MPG**: 32 **Top Speed:** 124 **0-60:** 9.5
Gearbox: Man/Auto **Boot:** 421 **Seats**: 4

Toyota Paseo 1996 - 1999

The utterly forgotten bland-coupe based on a Tercel saloon and ignored in Blighty because there were Tigras and Pumas to play with. Toyota did try with the bright yellow Galliano special edition, but otherwise it was the SiT that had a glass sunroof and electric mirrors and the Si that had alloys, CD and rear spoiler. Tough as a Tercel. That's it.

Engines: 1.5 **MPG**: 40.9 **Top Speed:** 114
0-60: 10.9 **Gearbox:** Man **Boot:** 277 **Seats**: 2+2

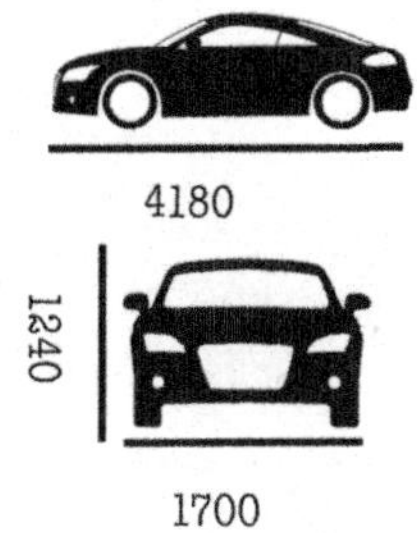

Toyota MR2 1990 - 1999

Bulked up, more hairdressy and softer as tin or targa top. Came with electric mirrors and alloy wheels, T-Bar had wipe down leather. For '92 cats, fresh alloys, new front and uprated suspension. From '94 high compression engine and side impact beams, better suspension and new rear lights. Toyota toughness, knobby owners neglect essentials.

Engines: 2.0 **MPG**: 35.8-35.9 **Top Speed:** 124-135
0-60: 7.2-9.3 **Gearbox:** Man/Auto **Boot:** 219 **Seats**: 2

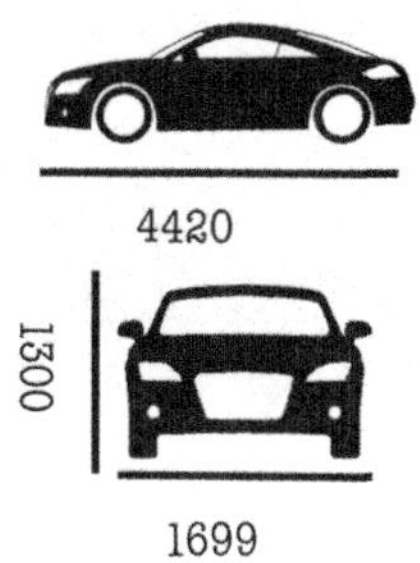

Toyota Celica 1990 - 1994

When four wheel drive was installed here's one of the most successful WRCs ever. For everyone else it was a soap shaped front wheel drive coupe that did not break down. In '92 had uprated cat engine, bigger alloys, discs and revised suspension. Finished off by abuse, so always best to find one of those cared for private owner examples.

Engines: 2.0 **MPG**: 30.1-35.6 **Top Speed:** 137-143
0-60: 6.7-8.2 **Gearbox:** Man **Boot:** 320 **Seats**: 4

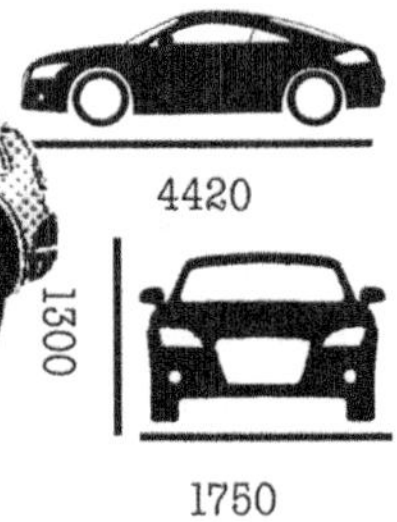

Toyota Celica 1994 - 1999

Goggle eyed creature seemingly aimed at the recently retired who might appreciate the large deep boot and inside it is pretty much like a Corolla. There is the GT Four though which is quite serious. In contrast from mid '95 the rather meh 1.8. The most overlooked Celica generation might mean a bargain still. Plenty have been experimented on.

Engines: 1.8 • 2.0 **MPG**: 27.2.-37.6 **Top Speed:** 124-153
0-60: 6.1-10 **Gearbox:** Man **Boot:** 283/593 **Seats**: 4

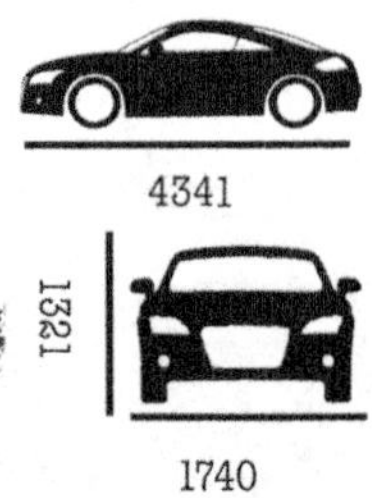

Toyota Celica 1999 2006

Seventh generation is a smart looking thing. Lightweight and sharp it came with cleverly engineered 1.8 engine in two different states of tune. Decent standard kit includes electric windows and allows, whilst the Premium added climate, leather and a CD. Compliant adults can get in the back. Many examples, not passengers, have been hammered.

Engines: 1.8 **MPG**: 33.6-36.7 **Top Speed:** 127-140
0-60: 7.2-8.7 **Gearbox:** Man **Boot:** 345 **Seats**: 4

Toyota Supra 1993 - 1996

Even more disappointing than a Paseo sales wise despite the full on supercar looks and performance. So cheap and available, with every extra known to humanity and that pram handle spoiler. On top there were dozens of grey import variants just to confuse everyone. Plenty to go expensively wrong and many modded to death.

Engines: 3.0 **MPG**: 25.5 **Top Speed:** 155 **0-60:** 4.9
Gearbox: Man/Auto **Boot:** 185/317 **Seats**: 2

Vauxhall Astra Coupe 1998- 2004

Stumpy yet stylish and this had some proper poke as well, especially the Turbo. Being a Vauxhall is good because parts are not a problem and relatively cheap. Being so closely related to the hatch means a big old boot and the split fold rears good take fully formed adults. Standard equipment is alloys, air and electric windows. Avoid the trashed ones.

Engines: 1.8 • 2.0 • 2.2 **MPG**: 31.7-36.2 **Top Speed:** 127-152
0-60: 7.0-9.5 **Gearbox:** Man/Auto **Boot:** 460/1150 **Seats**: 4

Vauxhall Tigra 1994 - 2001

Strange little 2+2 thing that wasn't exactly that sporty and hardly as practical as a Corsa, but appealed to the young, female buyer. Or hairdresser. Adequate 1.6 and less than adequate 1.4 petrols, but incredibly Lotus sharpened up the handling in 1997. Cheap to run. Rust, smokey engines, mostly worn out.

Engines: 1.4 • 1.6 **CO2:** 173-183g/km **MPG**: 39.0 **Top Speed:** 118-126 **0-62:** 9.8-10.5 **Gearbox:** Man **Boot:** 215/425

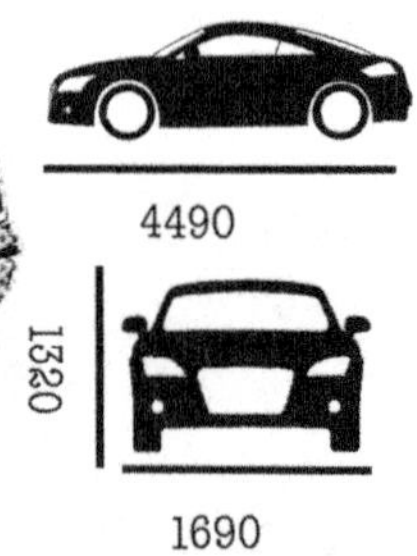

Vauxhall Calibra 1990 - 1998

The most slippery car you could buy at the time here was the nearest thing to the '90s Capri. Basic 16V best, four wheel drive, turbo and leathered up V6. Proper 4 seater too. Galvanised shells from '95 help, Cavalier mechanicals can at least be cheap, broken 4 x 4 gearbox isn't. Suspension usually shot, red top 16V head gaskets go.

Engines: 2.0 • 2.5 **MPG**: 30.9-35.5 **Top Speed:** 127-152 **0-60:** 6.4-9.8 **Gearbox:** Man/Auto **Boot:** 300/980

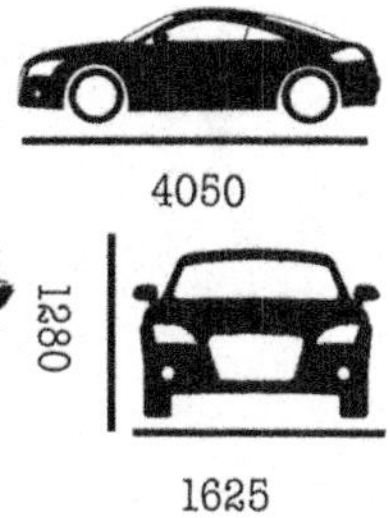

Volkswagen Scirocco 1981 - 1992

Underneath a Mark 1 with more rear headroom and luggage space. The Mk2 was not face lifted, but did gain body kits and various rear wings. Rot isn't so bad on these models, engines are tough and just get smokey and rattly with age. Gearbox linkage bushes go and become floppy. Electrics tend to be OK, this is solid Golf stuff and good value.

Engines: 1.6 • 1.8 **MPG**: 34.7 **Top Speed:** 120
0-60: 9 **Gearbox:** Man/Auto **Boot:** 346/920 **Seats**: 4

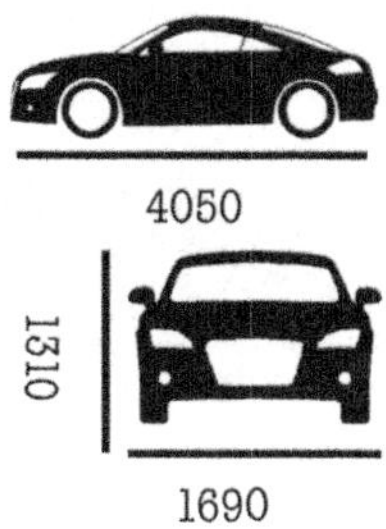

Volkswagen Corrado 1989 - 1995

A grown up Scirocco that contains plenty of Mark 2 Golf. Great to drive whatever the engine and there was a supercharged G60 and VR6. The Banger is the 2.0 8 valve, cheaper than Golf GTI. Decently rustproofed but some panels a problem now. Those engines are great provided not too tired and smokey. Suspension bushes, seized brake calipers.

Engines: 1.8 • 2.0 • 2.9 **MPG**: 28.3-35.5 **Top Speed:** 124-146
0-60: 6.7-10.6 **Gearbox:** Man/Auto **Boot:** 235/810 **Seats**: 4

Volvo 480 1987 - 1995

Dodgy Dutch build quality at the time didn't help. It's a 440-460 with less doors and was galvanised, but this hasn't helped much. Side impact beams from '93 and some interesting special editions. Wedge shape and pop up lights are good, everything else is fairly grim, only made an average banger, but the fanbase luv 'em, so buy from them.

Engines: 1.7 • 2.0 **MPG**: 34.2-36.3 **Top Speed:** 116-124
0-60: 8.8-10.6 **Boot**:161 **Gearbox:** Man/Auto

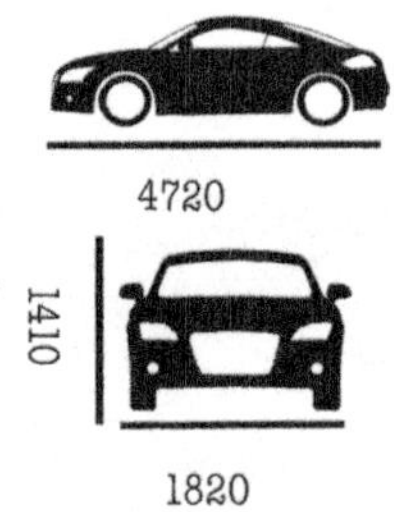

Volvo C70 1997 - 2005

Full on four seater with luggage space for all. Five cylinder engines some with turbos were good and eventually a cheaper to own 2.0 arrived. The value for money specification included air conditioning, electric most things and a decent stereo. Niggly problems with electrical relays sometimes. Kenny still reckons these are great.(back cover)

Engines: 2.0 • 2.4 • 2.5 **MPG**: 28.5-29.7 **Top Speed:** 124-155
0-60: 6.9-9.9 **Gearbox:** Man/Auto **Boot:** 371 **Seats**: 4

Notes.....

Bangervertibles

Same question as with the coupes, why on earth would you buy convertible car if you wanted to go Banging about? Sometimes it will be a matter of price and at other moments opportunity. Just because the roof is missing does not make it less of a Banger. There are so many bargains. It is a cliché to say buy when it is cold and winter, but occasionally that is very much the case. Desperate seller with a torn canvas hood meets keen Banger buyer.

There was a solid complement of can opened hatchbacks. Ford Escorts were always terribly average, the Vauxhall Astras was rather better and you could always rely on the Golf, especially the '70s original that continued to survive and thrive. The Renault 19 cabriolet was rather forgettable, but the Rover 200 quite a sharp little thing.

Four proper seats is underrated, luckily Mercedes supplied some very grown up ones and the original Saab 900 was as solid as a block. Otherwise the Vauxhall/Opel related ones are very stylish and great value. The Volvo C70 was a nice surprise and made a change.

Yet again the '90s was rich in creditable chop tops. Not everyone likes a Mazda MX-5 but it proved that the handy, affordable convertible was far from dead if it was well made and great to drive, unlike some of those '60s and '70s ones. Let's give a special mention to the Suzuki Cappuccino, because if the MX-5 was a properly made Lotus Elan, then the Cappuccino was a less leaky Midget and tighter handling Spitfire.

However, if you wanted to be reminded what open toppery used to be like there was always the Reliant Scimitar SS1. Not an especially bad car, but not great styling. Still bringing in the Alfa Romeo Spider was a wonderful throwback to a cooler era. The Fiat Barchetta was another reinterpretation that should be celebrated and bought. The '90s were great for Bangers with the roof missing..

Alfa Romeo Spider 1990 - 1993

Graduated to a new decade and that meant it was cleaned up with new plastic bumpers front and rear. More important was the fuel injected engine. It was possible to get a three speed automatic and the Veloce trim meant leather and air. Technically more of a Banger in the '80s, but these come as imports from the States and Europe. Rust, shabbiness.
Engines: 2.0 **MPG**: 30.7 **Top Speed:** 118 **0-60:** 9.4
Gearbox: ManAuto **Weight**: 1120 **Boot:** 300 **Seats**: 2

Alfa Romeo Spider 1996 - 2004

UK make do with the twin spark 2.0 litre, before V6 arrived. Quality roof lining shrouds the mechanism and the whole system neatly folds away beneath the metal roof cover. In 2003 there was a mild facelift and improved 2.0 JTS and 3.2 V6 engines. Caution: used Alfa. It's a proper risky Alfa buy. MOT failure level rust, smokey engine damaged roof.
Engines: 2.0 • 3.0 **MPG**: 24.1-30.7 **Top Speed:** 137-154
0-60: 8.9-9.6 **Gearbox:** Man/Auto **Boot:** 147 **Seats**: 2

Audi 80 Cabriolet 1992 - 2002

The '90s Cab with the Princess connection. Classic from day one, quality and overall durability meant a life of domestic Bangerdom awaited. Lots of engines. Power hood with the 2.8, removable hardtops from '97 on. Bringing back up to royal standard costs a lot. Engines suffer from camshaft wear, steering wear, suspension and rubbish paintwork.

Engines: 1.8 • 2.0 • 2.3 • 2.6 • 2.8 **Top Speed:** 121-137
MPG: 30.3-32.9 **Weight**: 1350 **Boot:** 230 **Seats:** 4

Audi TT Roadster 1999 - 2006

The choice is what state of tune for the 1.8T engine and whether or not Quattro four wheel drive is required. There was the option of a 3.2 V6 and S-Line spec option packs. The simple manual hood is easy to use and there is an electronic glass wind deflector which is useful. Ignition packs were not great. Also ECUs, brakes, bodywork and hood damage.

Engines: 1.8 • 3.2 **MPG**: 29.4-35.0 **Top Speed:** 133-147
0-60: 6.5-7.8 **Boot:** 220 **Seats**: 2

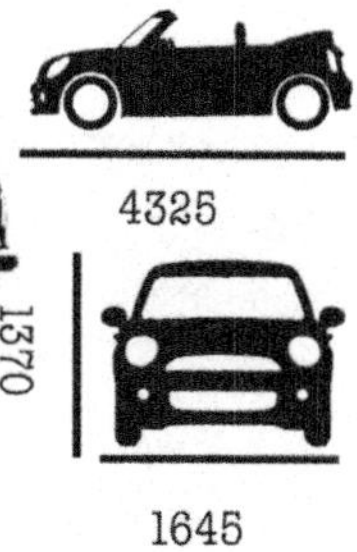

BMW 3 Series Convertible 1986 - 1993

'80s open topped legend continued into the '90s thanks to demand and epic build quality. Great 6 cylinder engines, the fours will plod on though. 318i from '91 plus restyled rear lights and body colour bumpers. Late model uprated specs Lux pack with cross spokes, metallic and spoilers. Electric hood and leather on 6 cylinders, Killed by rot, head gaskets.
Engines: 1.8•2.0•2.5 **MPG**: 30.1-35.3 **Gearbox:** Man/Auto **Boot:** 312 **Seats**: 4 **Weight:** 1270

BMW 3 Series Convertible 1994 - 1999

Not quite as E30 solid but probably slicker and certainly fine to own and drive. Usual slant of good fours, great sixes except the M52 with Nikasil bores, most sorted out by now. The fours are Banger tough Revamped in '96 a bit with fatter grille surrounds and slimmer wing indicators. Pretty reliable, but fluid leaks, iffy trim fragile, electrics go 'bong'.
Engines: 1.8•2.0•2.5•2.8• **MPG**: 31.0-35.8 **Top Speed**: 130-147 **0-60:** 7.3-9.9 **Gearbox:** Man/Auto **Boot:** 230

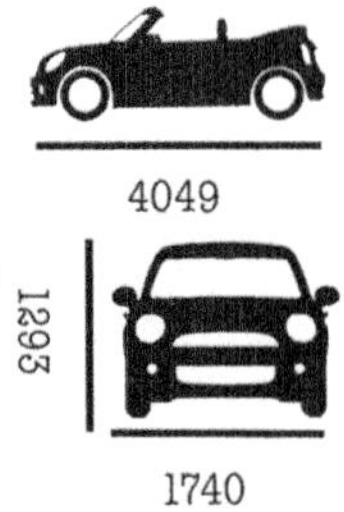

BMW Z3 1996 - 2002

Underneath it is a very old 3 Series effectively from the '80s so easy to live with and fix. Basic 1.9 petrol engine from the off followed by the 2.8 6 cylinder. Slightly face lifted in 1999 to look a bit more like the full on M Coupe. Certain models more 'classic' than others, proper Banger Bimmer with the roof missing. Bore wear, suspension bushes. **Engines:** 1.9•2.2•3.0•3.2 **MPG**: 25.4-36.2 **Top Speed: 122**-155 **0-60:** 5.5-10.4 **Gearbox:** Man/Auto **Boot:** 165 **Seats**: 2

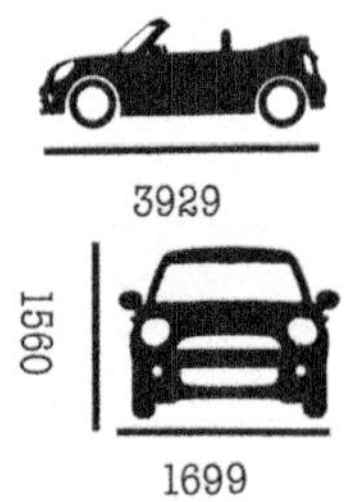

Citroen Pluriel 1994 - 2010

Effectively a Lego kit with lots of plastic parts that can be manually deconstructed into a convertible. Eventually. There is nowhere to store the bits as you move from sunroof to targa so hopefully it won't rain too much. A humble C3 with a decent diesel option. A cheap but hands on way of getting a convertible. Fragile. Electricals fail such as locks and windows. **Engines:** 1.4•1.6•1.4D **MPG**: 41.5-62.8 **Top Speed:** 98-117 **Boot:** 137/266 **Seats**: 4

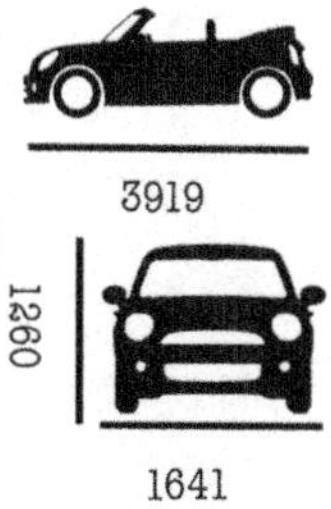

Fiat Barchetta 1995 - 2005

Wonderful characterful, left hand drive although there were aftermarket right hook conversions. Front wheel drive puts some purists off, but they are twits. Manual hood is far easier to operate than an old rubbish MGB plus it packs away under a metal tonneau cover. Slight revamp in 2003. Overall charming, but rust and Fiat related fragility can kill it. Check everything. **Engines:** 1.8 **MPG**: 33.2 **Top Speed:** 124 **0-60**: 8.9 **Gearbox:** Man **Boot:** 165 **Seats**: 2

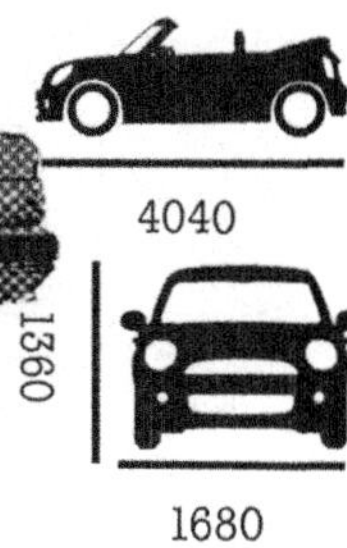

Ford Escort Cabriolet 1990 - 1995

Mk V Escort the Cabriolet model choice being either a 108bhp 1.6i or a 105bhp cat version. A 1.4-litre 60bhp Escort Dash Cabriolet followed in '92, few were fooled. Spring '93 Oval grille Facelift '93 with 130bhp 1.8-litre Si and 90bhp 1.6-litre Silhouette, Mistral and Solar editions. Electricals never great, ECUs and windows, noisy camshaft. **Engines:** 1.4 • 1.6 • 1.8 **MPG**: 25.6-37.8 **Gearbox:** Man
Top Speed: 103 -125 **0-60:** 13.7-10.1 **Boot:** 380 **Seats:** 4

Ford Escort Cabriolet 1995 - 1998

Mk VII or 'Escort 95'. With additional body strengthening. 130bhp 1.8-litre Si joined by a 115bhp. Calypso powered by the 90bhp 1.6-litre. Ghia versions followed in 1996, powered by the 115bhp 1.8-litre Zetec units. Automatics offered in either Calypso or Ghia trim, powered by a 90bhp version of the 1.8-litre Zetec.Blown head gaskets, bodywork rust. **Engines:** 1.6 • 1.8 **MPG**: 42.7-44.2 **Gearbox:** Man/Auto **Top Speed**: 103 -125 **0-60:** 11.6-10.1**Boot:** 380 **Seats**: 4

Honda S2000 1999- 2009

Incredibly well built and a rear wheel drive model in its own right, rather than being an Accord underneath. Engine is a work of engineering genius, a hugely powerful four cylinder 2.0 litre. Almost as quick is the electric roof which takes just 10 seconds to do its stuff. It's a genuine classic. Some rust, not cheap to fix if engine shot. **Engines:** 2.0 **MPG**: 28.2 **Top Speed:** 150 **0-60**: 6.2 **Gearbox:** Man **Boot:** 143 **Seats**: 2

Jaguar XJS Convertible 1991 - 1996

These revised '91s have galvanised bodies so not to be confused with the old rotters. Often full of problems, Despite simplified construction they still oxidise. Revised AJ6 engine is better and became the AJ16 in '94. Low miles can be fatal, they need exercise, so smoke, coolant, oil leaks, worn suspension and radius arms. Lecky roof breaks.
Engines: 4.0 • 5.3 • 6.0 **MPG**: 18.8-24.3 **Top Speed:** 141-161
0-60: 6.6-7.8 **Gearbox:** Auto **Boot:** 393 **Seats**: 2+2

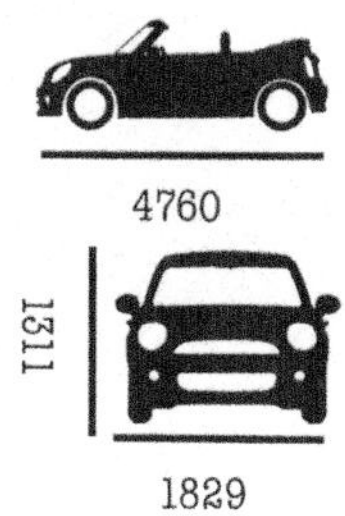

Jaguar XK8 1996 - 2005

Still related to the previous XJS, 2+2 tight in the back, the classic reason to buy is that it is a sheer value for money Jag. Comes with an electric roof which still needs to be manually packed away beneath the tonneau cover and that robs the boot of space. The V8 is thirsty but good. Needs proper, costly, looking after. Electrics, ECUs, rust, suspension. **Engines:** 4.0 • 4.2 **MPG**: 22.2-24.1
Top Speed: 155 **0-60:** 4.4-7.1 **Gearbox:** Auto **Boot:** 306

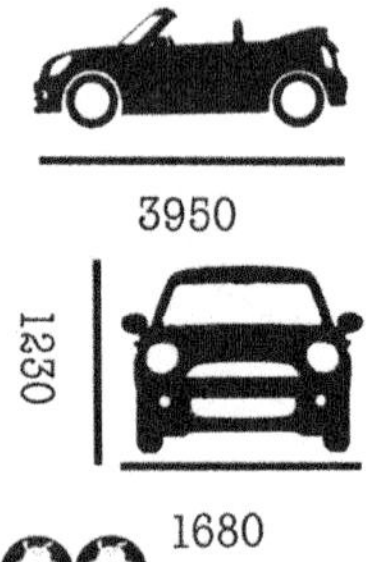

Mazda MX-5 1989 -1998

Pop up lights and 1.6, or 1.8 engine are all the ingredients you need for a fun time. Easy to live with and a very simple manual hood. There are Japanese imports with extra kit like air con and much less underseal. Overall, best Banger chop top in history. Rust, brakes, hood, nothing unfixable.

Engines: 1.6 • 1.8 **MPG**: 31.7-36.2 **Top Speed:** 118-127
0-60:8.0-10.6 **Gearbox:** Man **Boot:** 160 **Seats**: 2

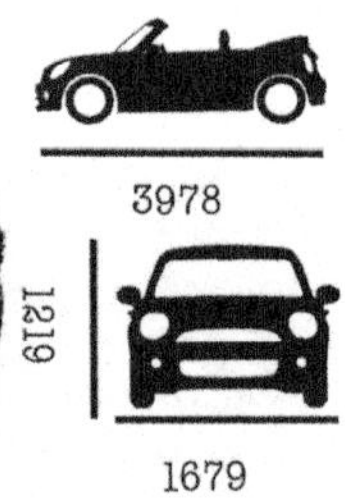

Mazda MX-5 1998 - 2005

All new MX-5 without pop up lights and with a (non contact) 1.6, or 1.8 engine. Easy to live with and a simple manual hood. There are still Japanese imports with extra kit like air con and less underseal. It is a cliché but this really is the best convertible in history. More special editions than Talbot (ask your Grandad). Rust, brakes, hood, nothing unfixable.

Engines: 1.6 • 1.8 **MPG**: 31.7-36.2 **Top Speed:** 119-120
0-60:8.9-9.7 **Gearbox:** Man **Boot:** 144 **Seats**: 2

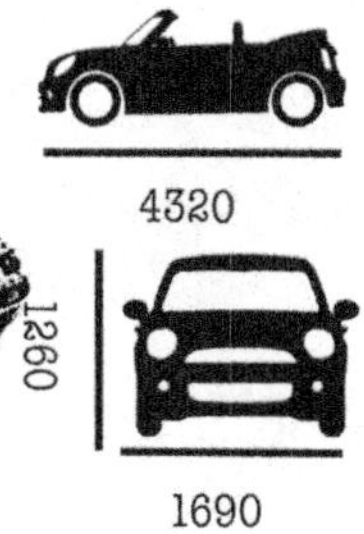

Mazda RX-7 Convertible 1989 - 1992

Largely forgotten, the Rotary was unique and perfectly suited to cruising about. Despite being well built, the Rotary engine is a harbinger of oil leaks and expensive doom sometimes with an added turbo. Excessively clever when compared to an XJS or SL, the massive depreciation helped but the spinning pistons ultimately didn't.

Engines: 1.3 **MPG**: 23 **Top Speed:** 149 **0-60:** 6.7
Gearbox: Man **Boot:** 200 **Seats**: 2

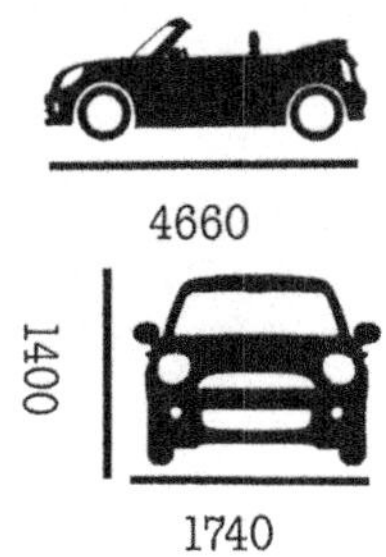

Mercedes CE / E Convertible 1987-1997

Accountant's idea of a Cabriolet in W124 wrapping. From August '93 badged as Es with mild facelift and spec upgrade for 1995 which meant walnut trim, hole alloys and headlamp wash/wipe. Model run outs in '96 got cruise and anti-skid. Other worldly build quality, but not immune from early ECU failure and shonky auto boxes

Engines: 2.2 • 2.3 • 3.0 • 3.2 **MPG**: 25.3-33.3 **Top Speed:** 126-147 **0-60:** 7.5-10.6-**Gearbox:** Man/Auto **Boot:** 480 **Seats**: 4

Mercedes SL 1989 - 2002

Nomenclature nerds know its an R129 and one the last truly great Merc. Electronically controlled 722.6 automatic gearbox in 1994 is rather complicated. Biggest facelift in 1998, 17" wheels, new taillights, bumpers, and new seats, chromed dials. Under the bonnet less characterful V6 and V8. Fragile wiring looms fairly terrifying.

Engine: 2.8 • 3.0 • 3.2 • 5.0 **MPG**: 22-25.1 **Top Speed:** 140-155 **0-60:** 7.2 **Gearbox:** Auto **Boot:** 260 **Seats**: 2

Mercedes CLK Convertible 1998 - 2003

Proper 4-seat family sized cab with a colossal range of petrol engines from adequate to overkill. Three layered hood is electrically operated and manually released from the driver's seat. Although it gets complicated there are the supercharged 230K and 320 and 430 V6 are Sport, Elegance and Avantgarde trim. MOT failure level rust.

Engines: 2.0 • 2.3 • 3.2 • 4.3 **MPG**: 24.2-28.8 **Top Speed:** 129-155 **0-60:** 5.4-9.7 **Gearbox:** Auto **Boot:** 419

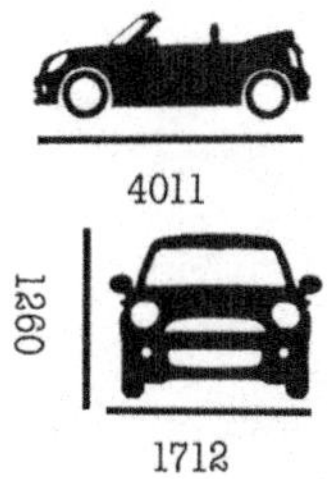

Mercedes SLK 1996 - 2004

An SL for the masses. Available with little 2.0 and 2.3 supercharged engines at first, then a 3.2 V6. Just the two metal panels in that roof which takes around 25 seconds to fold away. Comes with leather and most electric things. Long in the tooth and roof. Caution needed, some serious rust, roof can be knackered.

Engines: 2.0 • 2.3 • 3.2 **MPG**: 24.8-28.5 **Top Speed:** 138-142 **0-60**: 5.2-8.2 **Gearbox:** Auto **Boot:** 345/144 **Seats**: 2

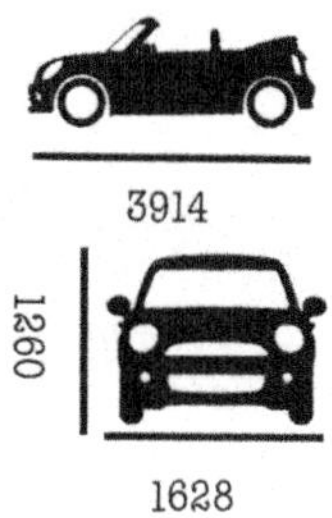

MGF 1995 - 2002

Unexpectedly rather good when it rolled up with two versions of the 1.8 engine, which was positioned behind the seats. Revisions in 1999 meant a mild restyle, a Steptronic gearbox and electronic steering. Five models from 2001 with 1.6 models. Beware blown head gaskets and that Hydragas suspension could give owners sleepless nights.

Engines: 1.6 • 1.8 **MPG**: 36.3-38.4 **Top Speed:** 112-130 **0-60:** 7.0-9.3 **Gearbox:** Man **Boot:** 210 **Seats**: 2

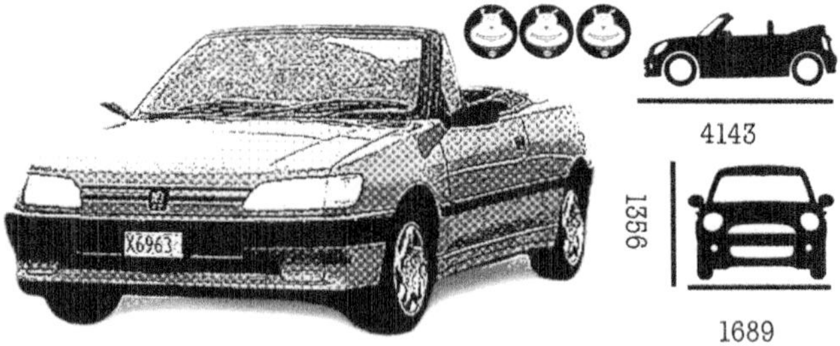

Peugeot 306 Cabriolet 1994 - 2002

That hood was well insulated for the time and electric powered. Simply unclip and push the button and it disappears into the raised cover in front of the boot. The optional hardtop is worth having as it is so stylish. Facelifted in 1997 and it got ABS brakes. Most are worn out or scrapped sadly. Everything could be broken or rusty.

Engines: 1.8 **MPG**: 30.7 **Top Speed:** 117 **0-60:** 12.2 **Gearbox:** Man/Auto **Boot:** 274 **Seats**: 2+2

Porsche Boxster 1996 - 2004

This is the entry level sporting Porker, a cut price 911 cabriolet, but with more luggage space in the front boot. Had the same sort of flat air cooled engines, so sounds like a Porsche, in 2.7 and 3.2 litre sizes. Major engine issues from cracked blocks to crankshafts, poor engine access part of the issue. Porsche running costs, but otherwise worth a go.

Engines: 2.7 • 3.2 **MPG**: 26.9-29.1 **Top Speed:** 157-164 **0-60**: 5.7-6.4 **Gearbox:** Man/Auto **Boot:** 260

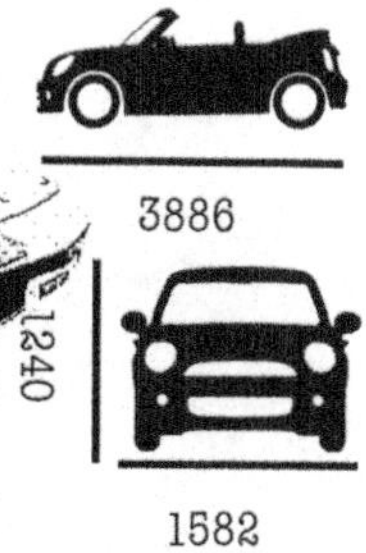

Reliant Scimitar SS1-SST 1984 - 1995

British roadster revival should have started here, but fit , finish and fugliness too terrible. SST Turbo might have fixed that in '90. Didn't. Ford and Nissan engines a good thing of course. Quite a Brit Bitza, with suspension from a Sierra and seats from a TR7. Oh yes and it became the Scimitar Sabre with 1.4 K-Series. Looked less scary but a lost cause.

Engines: 1.3 • 1.4 • 1.6 • 1.8 **MPG**: 33.2-34.9 **Top Speed:** 99 - 111**0-60:** 9.6-12.7 **Gearbox:** Man **Boot:** 190 **Weight:** 840+

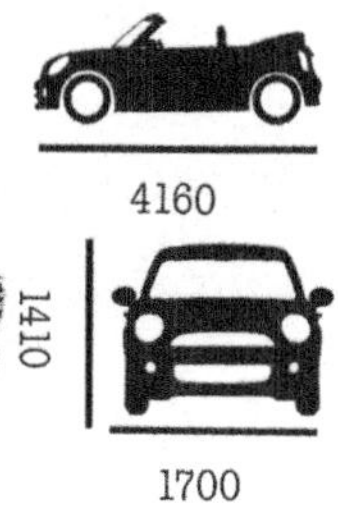

Renault 19 Cabriolet 1992 - 1996

'Phase II' 19s with a new corporate front grille and smoked tail lights and the cab had the 1.8 and 1.8 16v. Then 'Phase III' in September 1994, non-16v 1.8 models now dropped but had side impact beams and power steering and plusher models had driver airbags. There are hardy any in circulation, survivors are electrically challenged.

Engines: 1.8 **MPG**: 34 **Top Speed:** 129 **0-60:** 9.4
Gearbox: Man **Boot:** 255/360 **Weight**: 1182 **Seats**: 4

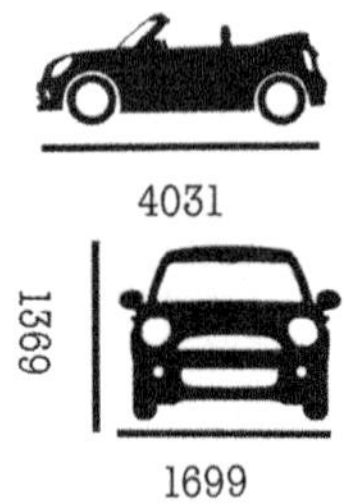

Renault Megane 1997 - 2003

That fabric roof is though easy to operate by unclipping and twisting a handle on the screen then pressing a button followed by a gentle 25 second folding procedure. Got a new nose for '99 and a restyle, plus some airbags and the potent 2.0 IDE engine. Gearbox and electrics. Clutch repair difficult, coil packs +flywheel sensors.

Engines: 1.6 • 2.0 **MPG**: 37.2-40.4 **Top Speed:** 124-131
0-60: 9.0-10.1 **Gearbox:** Man/Auto **Boot:** 280 **Seats**: 4

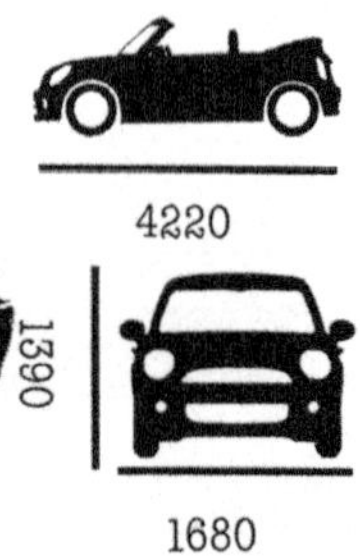

Rover 200 Cabriolet 1992- 1999

Carbon date the running gear to '80s Honda Concerto. That didn't make it any less of a decent chop top. When the new 25 came out in '95 it lost the 200, got the small K series engine and a CVT gearbox. Standard and SE were the buyer options. The Honda 1.6 is solid, but costs a bit more to look after. The K series can blow a gasket if not looked after.

Engines: 1.4 • 1.6 **MPG**: 32.3-41.7 **Top Speed:** 106-118
0-60: 10.5-6.2 **Gearbox:** Man/Auto **Boot:** 300 **Seats**: 4

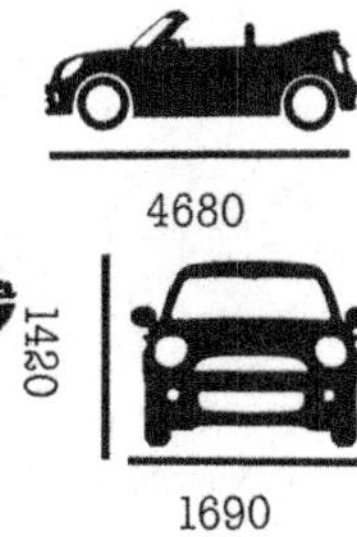

Saab 900 Cabriolet 1984 - 1993

Ancient underpinnings, but a rock solid Banger legend, that never needed a roll-bar. Light Pressure Turbos in 1990, Limited run models to '93 SE and Turbos. 16 valvers are fun. Six figure mileages are a given with these. Non Turbo may be underwhelming, but less Banger bother. Rattling timing chains, noisy camshaft, blowing turbos. Brilliant.

Engines: 2.0 • 2.0 Turbo **MPG**: 27.9-28.8
Gearbox: Man/Auto **Boot:** 650 **Seats**: 4

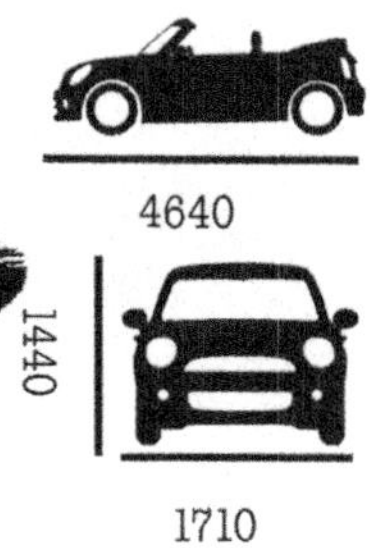

Saab 900 Cabriolet 1994 - 1998

Four seater and still looks sensationally modern. There can be lots of hood related issues that frustrate owners. Engines were either 2.0 or 2.3-litre four cylinders (and a 2.0-litre turbo) or Vauxhall-2.5-litre V6. '95 colour-coded bumpers '97 better front seats better gear-change, larger brake discs. .Oil leaks, worn timing chains, coil packs.

Engines: 2.0 • 2.3 • 2.5 **MPG**: 31.4-32.1 **Top Speed:** 134-140
0-60:6.9-8.0 **Gearbox:** Man/Auto **Boot:** 378

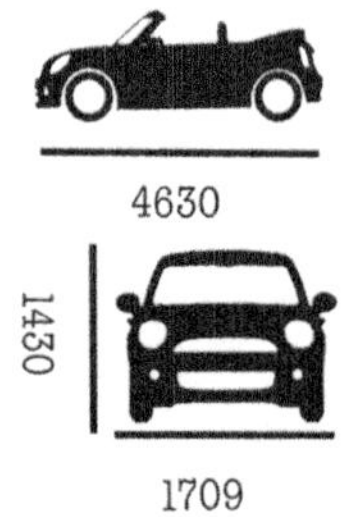

Saab 9-3 Cabriolet 1998 - 2003

Fantastic seats and always sensational value. Lots of criticism about the hard ride in town and the floppiness when being driven enthusiastically. Nobody cares, it is a classy four seater. Best not bought as any sort of project. Parts getting expensive. Oil leaks, worn timing chains, coil packs don't last long. **Engines:** 2.0 • 2.3 **MPG**: 31.4-32.1 **Top Speed:** 134-146 **0-60**:6.9-8.0 **Boot:** 294

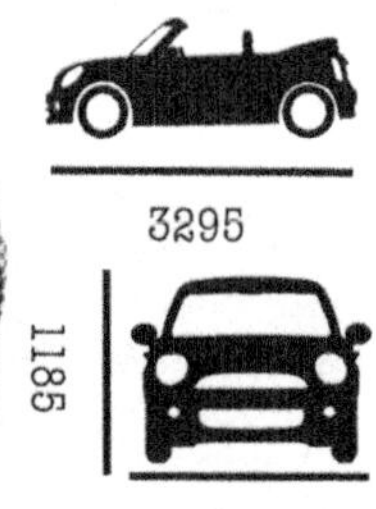

Suzuki Cappuccino 1993 - 1996

It is like they copied the MGF and made it better, apart from the luggage space, because there isn't any, or very much. More importantly putting the hood down is a flick of the wrist from the driver's seat is a doddle, plus it has a proper heated glass window. Even has air con, great if you have the optional hardtop. Reliable and will take a ton of abuse.
Engines: 1.8 **MPG**: 38.2 **Top Speed:** 130 **0-60:** 7.2
Gearbox: Man **Boot:** 90 **Seats**: 2

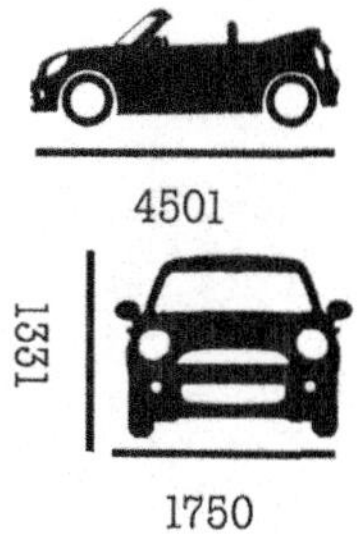

Toyota Celica Convertible 1994 - 1999

Taking the roof off often improves some clunky coupes, but not here. It cost a bomb so no one bought them. Not much happened except in 1995 it got a high level brake light. It's a Toyota so there are no real worries, apart from grey imports that never much in the way of rust proofing. Essentially a two door chop top Corolla and just as competent.

Engines: 2.0 **MPG**: 27.2 **Top Speed:** 134
0-60: 8.5 **Gearbox:** Man **Boot:** 283 **Seats**: 4

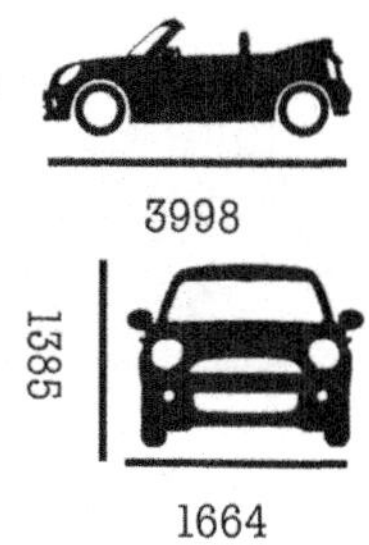

Vauxhall Astra Convertible 1987 - 1993

Quite a lot of ordinary enough Mark 1 underneath the Bertone-built convertible with its standard power-operated hood was excellent value for money. Dumpy to look at and drive, plus everyone wants and wanted a GTE, but the basic 1.6 does a decent open air job. Most have rusted away, smokey rattly and overdue for a cambelt change.

Engines: 1.6 • 1.8 • 2.0 **MPG**:34.1-42.2 **Top Speed:** 106-121
0-60: 10.5-9 **Gearbox:** Man/Auto **Boot:** 289 **Seats**: 4

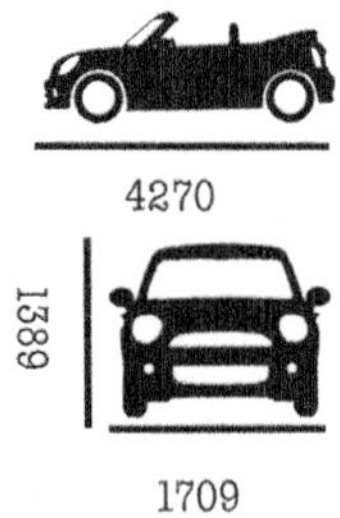

Vauxhall Astra Convertible 1993 - 1999

Blunt, basic and effective convertible 4 people. There is room in the back. They can be protected from the elements by the three layer hood which can even be operated by the key fob (high tech at the time) and takes 30 seconds to pack away. Can be had with the 2.0 Turbo. Corroded ECUs, cockpit leaks, failing Air conditioning and water pumps.

Engines: 1.6 • 1.8 • 2.0 • 2.2 **MPG**:34.4-42.2 **Top Speed:** 118-149 **0-60:** 7.5-12.0 **Gearbox:** Man/Auto **Boot:** 330 **Seats**: 4

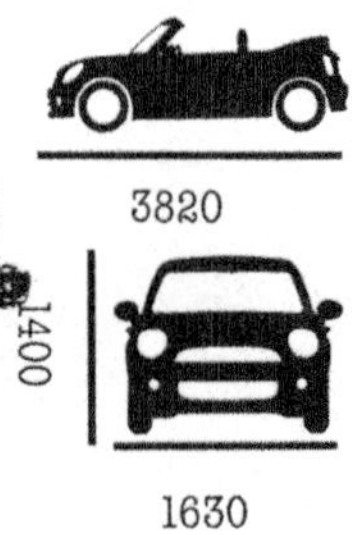

Volkswagen Golf Cabriolet 1987 - 1993

Ancient old thing that remained ultra desirable. From 1990 an electronic hood became standard all had colour coded body kit including basic cheery 1.5 Clipper Then in '91 GTI Sportline and Rivage, with sexy Recaro, BBS extras, Clipper got electric windows. Had same buying issues (bore wear) as original Mark 1, with added leaks, but nobody cared.

Engines: 1.8 **MPG**: 33.3-36.5 **Top Speed:** 108-111 **0-60:** 9.4 **Gearbox:** Man/Auto **Boot:** 220 **Seats**: 4

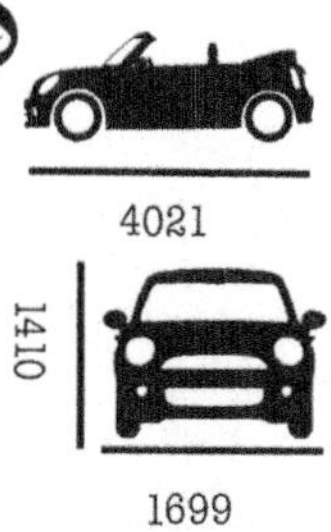

Volkswagen Golf Cabriolet 1994 - 2003

Mark 3 Golf was never the best.. Folding rear seats boosted the boot area and apart from the basic model, the hood was electrically operated. Just the petrols, a 1.8, 1.6 and 2.0 in the noughties are all just adequate, but certainly reliable and should last forever. 1998 Mark 4 front end, but not interior. Feel cheated? Rust, bore wear.

Engines: 2.0 **CO2:** 194g/km **MPG**: 34.9 **Top Speed:** 119 **0-60:** 10.7 **Gearbox:** Man/Auto **Boot:** 269/447 **Seats**: 4

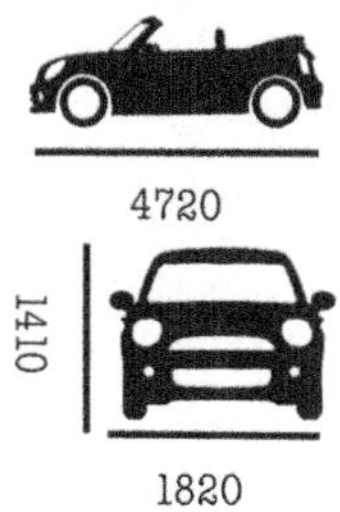

Volvo C70 1997- 2006

One of the most classy cabriolets of the '90s with advanced electric roofage which saw just one press of the button to make it all work, plus had a proper glass window. Incredibly four roof colours available to complement the bodywork. A ton of interesting engines from a basic 2.0 right up to the superconvertible T5. Rust, electrics, brakes and suspension.

Engines: 2.0 • 2.3 • 2.4 **MPG**:28.5-29.7 **Top Speed:** 130-146 **0-60:** 7.5-9.9 **Gearbox:** Man/Auto **Boot:** 201

LuxoBangerBarge

Great big comfy cars are truly wonderful things. They also make great Bangers. That is because depreciation is very brutal to these Barges. Those daft enough to buy new, or at least clever enough to offset some company cash, leave behind a depreciation time bomb that we can easily diffuse.

The 1990s serves up a decent collection of blinged up saloons which in most cases come from the usual German sources. There are some magnificent examples of the art exemplified by the BMW 7 Series. Sensational to drive and in E38 form in particular, exceptionally well put together. Those six cylinder engines are efficient and quiet whilst the electronics seem solid enough.

Less interesting to drive but possibly the supreme example of how to make a barge is the early '90s era Mercedes S-Class. For some it was too much tech and over indulgence, but the standard of fit, finish and reliability has been exceptional. Great limousine whether you are in the front or back.

The Audi A8 came of age, not much in the way of personality, but the Quattroness of it all meant there was no wallowing around. Build quality was incredible. Opposition to the Germans came in the shape of Lexus and the LS series is the cleverest of creations. A luxury brand from ground zero. Honda did not bother Acura-ising their Legend and perhaps they should have done.

Meanwhile the old guard, Rolls Royce and Bentley was learning to survive by updating their ancient architecture and V8 for a generation that expected so much more. They traded on their legacy badges before succumbing to the inevitable German uber juggernauts, BMW and Volkswagen. Luxury now has a German accent.

At least we still had Jaguar, oh they were now American and going by the Cadillac Seville, didn't have the faintest idea of how to pamper. At least these Jags are cheap, nicely finished and comfy. Run one till it breaks down and chuck away. It's the only way to Barge without going mad or bankrupt.

Alfa Romeo 164 1988 - 1997
Surprisingly accomplished big Alfa. Reassuringly odd interior and controls. Easily the sexiest of the Type 4 cars (Croma/Thema/9000) which the Bangerpedia promotes to luxury status and although the FWD+V6 combo a handful, one of the best big saloons to drive hard. Always needed looking after, electrics, suspension and V6 need attention.
Engines: 2.0 • 3.0 **MPG**: 27.1-36.1 **Speed**:152 **0-60:** 9.9-7.7
Boot: 505 **Weight**: 1200 **Seats**: 5 **Doors:** 4

Alfa Romeo 166 1999 - 2005
Inside the dashboard looks lovely, all overlapping dials, timber rimmed steering wheel and plush leather on the Lusso models. It is a big car that likes going quickly and you can take corners very smartly indeed. Suspension bushes, oil pump failure, the usual electricals and needs timing belts sharpish. Sherlock might have a problem finding one now.
Engines: 2.0 • 3.0 • 3.2 **MPG**: 22.6-29.1 **Gearbox:** Man/Auto
Boot: 490 **Seats**: 5 **Doors:** 4

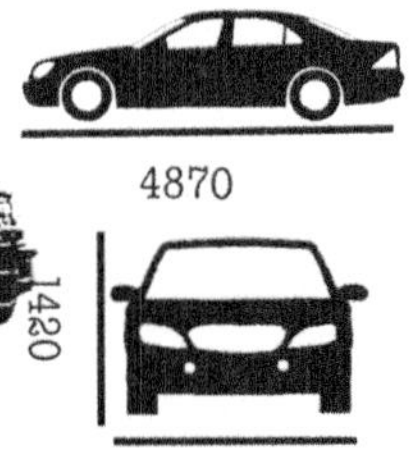

Audi 100 V8 1988 - 1993

Looks like a 100, but it shares only the doors with that model here is the blueprint for the great big, four wheel drive, hyper luxury uber saloon. Actually far too much tech for the time and not as dependable as it should have been. Historically important, but not worth the effort of tracking one down. A8 for effort, but C- for execution.

Engines: 3.6•4.2 **MPG**: 23 **0-60:** 6.8 -7.6 **Top Speed**: 152-155 **Boot:** 570 **Seats**: 5 **Doors:** 4

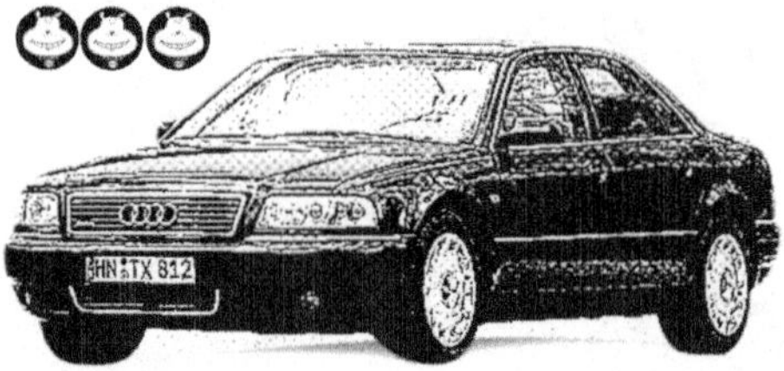

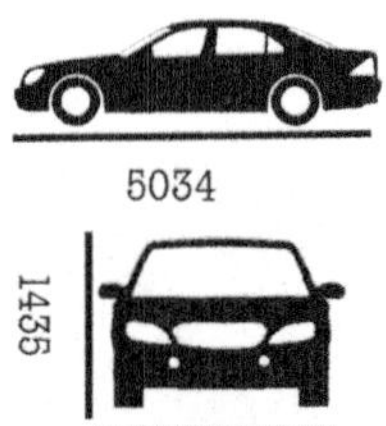

Audi A8 1994 - 2002

It is a Xerox'd up A6. A big anonymous saloon though given some character courtesy of the S8. The aluminum body once dented, costs a bomb to fix, so you wouldn't bother. V8 engines are great, Quattro makes it a super snowplough. Suspension, brakes may need sorting, automatic gearbox , air con and general electrics/ECUs.

Engines: 2.8•3.7•4.2 **MPG**: 20.2-26.4 **Boot:** 524 **MPH:** 135-155 **0-60:** 5.5-11.2 **Seats**: 5 **Doors:** 4

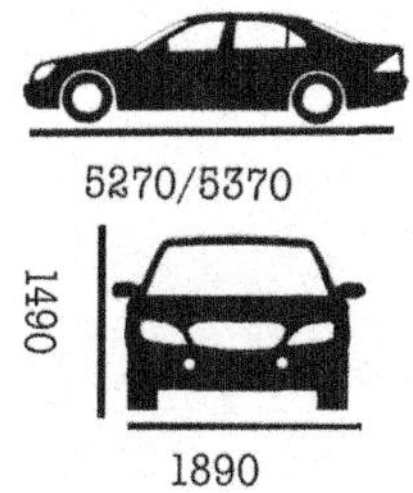

Bentley Brooklands 1992 - 1998

Moving on from the Mulsanne with useful ABS. 1993, 250bhp, then in 1996 a 300bhp light-pressure turbo, a Brooklands R in '98 the Turbo R's suspension and 17-inch wheels. Essentially entry level Bentley. Engine and GM gearbox bulletproof. Body can degrade. Turbos abused and everything else ex-hire neglected. **Engines:** 6.75 **MPG**:15 **0-60**: 9.8 15 **BHP**: 234/250 **Speed**:134 **Boot:** 550
Weight: 2430 **Seats**: 5 **Doors:** 4

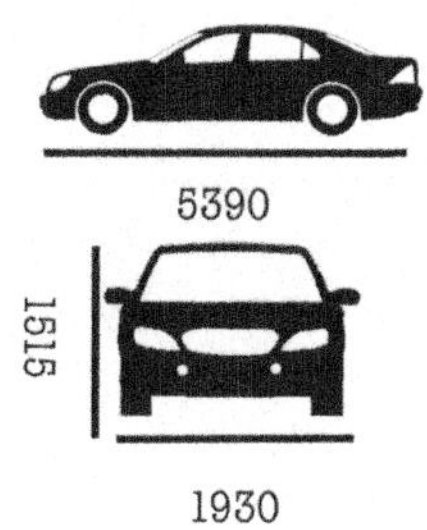

Bentley Arnage 1998 - 2009

The Bentley T, Roller Shadow of the modern era. It is dead smart and destined for wedding work. RL is the longer version. originally Green and Red Labels replaced by R and T models. Everything costs a bomb, but a lot of BMW bits in the early ones, so you need to be clever. Brakes, suspension and the complicated electrics. Cosmetics kills asking price.
Engines: 6.75 **CO2:** 495g/km **MPG**: 13.7 **Boot:** 374 **Seats**: 5 **Doors:** 4

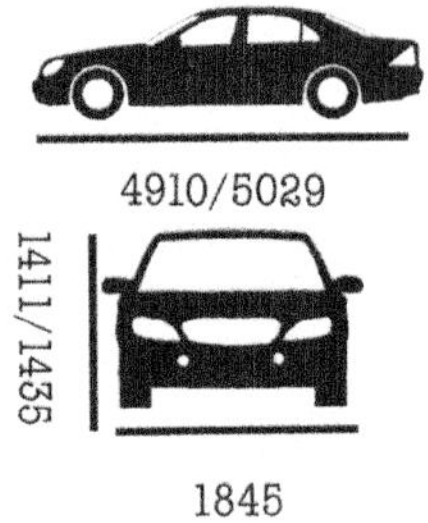

BMW 7 Series 1986 - 1994

BMLimoW comes of age with the E32. Revamped in '94 with new 730i and 740i with cruise, alloys and electric front seats. Lots of acronyms over the years including ASC+T and DSCIII which are great when they work. Overall pretty damn reliable and it would be the electronics which cause the most upsets, otherwise worn suspension, exhaust and brakes.

Engines: 2.8•3.0•3.5•4.0 **MPG**: 21.7-25.3 **Boot:** 500 **MPH**:140-155 **0-60:** 7.3-9.5 **Weight:** 1600 **Seats**: 5 **Doors:** 4

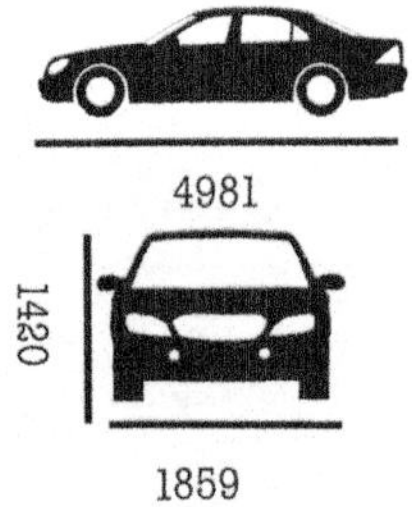

BMW 7 Series 1994 - 2002

As fun to drive as a little 3, but with lots more kit and comfort. The engine should go on forever, it is the suspension and brakes where the big expenses will be. V12 possibly an engine too far the 6 and V8 cylinders are the pick of the units. They don't come any bigger, better or harder. Buy one.

Engines: 2.8•3.5•4.4•5.4 **CO2:** 267-332g/km **MPG**: 20.5-25.4 **Boot:** 501 **Seats**: 5 **Doors:** 4

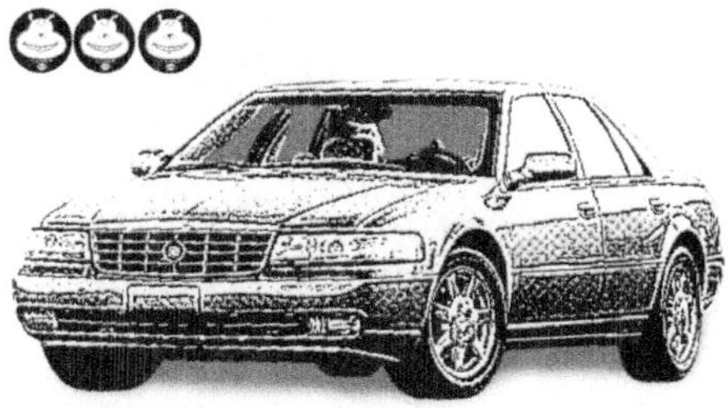

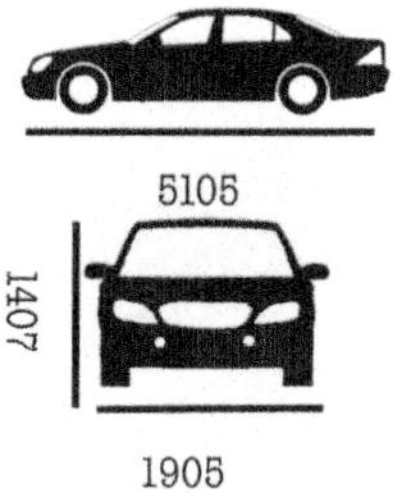

Cadillac Seville 1998 - 2002

For a time you could buy yourself a right hand drive Caddy with a 4.6 V8 available at selected Vauxhall dealers. Front wheel drive as well. Parts not cheap and it may be worth taking a transatlantic holiday to get bits, Not that many around, so it is rather exclusive. Just not as classy as the Germans. Proper Yank Tank. Oil leaks, coolant leaks, rusty bodywork and suspension. Electrics. **Engines:** 4.5 **CO2:** 330g/km **MPG**: 19.9 **Boot:** 445 **Seats**: 5 **Doors:** 4

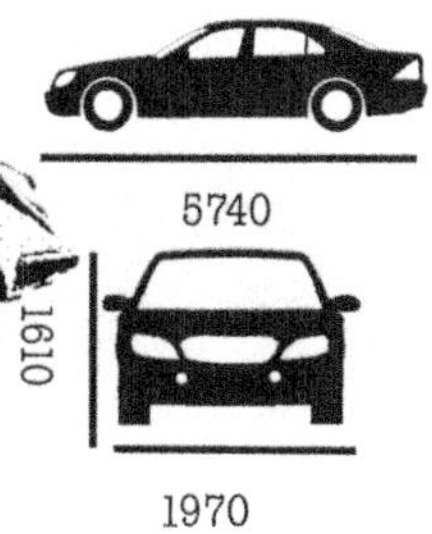

Daimler DS420 1968 - 1992

It's a posh people carrier for 7 passengers and their luggage. Beloved of Mayors, undertakers and royalty. It's a unique Daimler, built by Jaguar with an XK engine. It's got a great big bonnet, a chassis and is relatively straightforward to live with. Became and largely remains cheap as upmarket chips, used for Banger racing, or wedding duties in perpetuity. **Engines:** 4.2 **MPG**: 22.4-23.7 **Speed**: 110 **Weight:**2140 **Boot:** Not revealed... **Seats**: 6-8

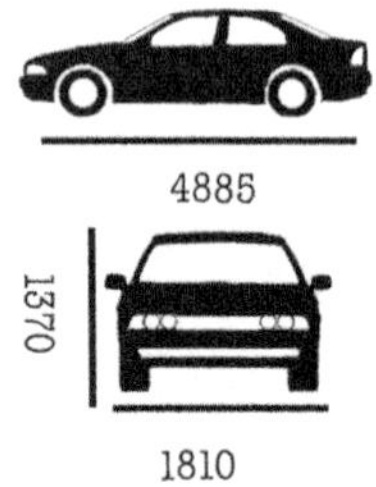

Honda Legend 1991 - 2004

It isn't really a Legend but it is a Honda. A great big comfy saloon that won't break down provided you look after it. This motor delivered ABS brakes and driver, then passenger airbags before they were widely available. A scaled up Accord and just as dull to look at and drive. Heavy use items like brakes and suspension take a bashing, also electrics.
Engines: 3.5 **CO2:** 291g/km **MPG**: 23.2 **Gearbox:** Auto **Boot:** 450 **Seats**: 5 **Doors:** 4

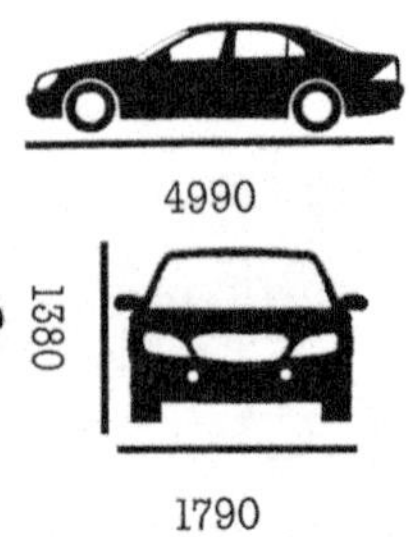

Jaguar XJ40 1986 - 1994

'80s slightly squarial Jag and obviously there are Daimler branded ones with fluted grilles and added walnut/chrome frippery. Very late launch of V12 is silent and deadly to your wallet in '93 along with ABS and driver's airbag. Not the most loved incarnation with irritatingly digital dashboard complicated and prone to failure.
Engines: 3.2 • 4.0 • 6.0 **MPG**: 18.4-27.2 **MPH**: 132-155
0-60: 6.8-8.5 **Boot:** 430 **Seats**: 5

Jaguar XJ12 Daimler Double Six 1989 - 1993

Sexy old shape retained for the big Vee. A big and beautiful creation with a ruinously expensive drinking habits. 1990 ABS, CD system in the boot. Jaguar Sovereign HE now called V12. Probably the best built of these beasts. Least loved for some time but properly rare and the very cheapest of any multi-cylinder saloon in captivity.

Engines: 5.3 **MPG**: 18.4 **Weight**: 1926 **Boot:** 400 **Speed**:143
0-62: 7.2 **Seats**: 5

Jaguar XJ (X308) 1994 -2003

Lovable old school Jag with a V8 from '97. The XJR super saloon cheap as old chips as well. Fairly snug inside if you like that sort of thing and there's a long wheelbase for work. Worn brakes and suspension, cracked windscreens, rattly timing chains. Blown head gaskets. Rust MOT failure. Executive worth finding and Super V8 have extra features.

Engines: 3.2 • 4.0 • 6.0 **MPG**: 22.4-26.9 **Speed:** 136-155
0-60: 5.3-8.9 **Boot:** 408

Lexus LS400 1990 - 1995

The greatest single luxury car invention of the '90s. A brand from nowhere, well out the back of the Toyotaskunkworks shed that shamed the existing German barge orthodoxy. ABS and twin airbags from '93. Suspension uprated as well, otherworldly sound system. Essentially a mega Corolla and just as reliable. Fixing things costs a bomb.

Engine: 4.0 **BHP:** 260 **Speed**: 155 **0-60**: 6.9 **MPG**:23
Boot: 430 **Weight:** 1796 **Seats**: 5 **Doors:** 4

Lexus LS400 1995 - 2000

Already perfect grown up luxury saloon polished to blinding brilliance, or something. Engine uprated, delivering more of everything, power, efficiency and overall waftiness. wheelbase longer so even more room in the back. Soundproofing goes up another level, yet the whole barge was 95kg lighter. Best of the best '90s Barges.

Engine: 4.0 **BHP:** 260 **Speed**: 155 **0-60**: 6.9 **MPG**:23 **Boot:** 430 **Weight:** 1796 **Seats**: 5 **Doors:** 4

5019/5136

1438

1821

Mercedes S-Class 1979 - 1991

The greatest plutocrat shifter of all time made it into the '90s and did so with the high tech kit others were just catching up. So airbags, ASR anti-slip and offset crash friendly bodywork. Those brilliant all-alloy V8 engines and ambience of utter solidity was a wonderful combination. Never really went wrong, even after being million mile taxis.

Engines: 2.8 • 3.0 • 3.9 • 4.2 • 5.0 • 5.6 **MPG**:17.8-22.5 **Speed**: 124 - 155 **0-60:** 7.1-10.9 **Boot:** 510 **Seats**: 5

5110/5213

1490

1890

Mercedes S-Class 1991 - 1999

The Grossest Limo of all, but arrived at the wrong time. Delightfully over-engineered with self-closing bank vault doors, double glazing and every gadget known to humankind. Revamped in '94 with more efficient engines and more spec brake assist, park sensors, an auto-pilot navigation system, adaptive damping. A problem if broken.

Engines: 3.2 • 3.2 • 4.2 • 5.0 • 6.0 **MPG**: 17.8-22.5 **MPH:** 130-155 **0-60:** 10.7-7.1 **Boot:** 525 **Seats**: 5 **Doors:** 4

Mercedes S-Class 1999 - 2005

This was a difficult time for Stuttgart, but the S-Class was better than a pretty ordinary bunch. Needs to be looked after otherwise it deteriorates rapidly. Most examples seem to be the 320 diesel. Longer models are available for private hire. V8 petrols and good standard kit. Oil and coolant leaks, bodywork and suspension corrosion. ECU and electrical failure. **Engines:** 2.8•3.7•4.3•5.0•5.5•3.2D **CO2:** 204-355g/km **MPG**: 19.1-36.7 **Boot:** 560 **Seats**: 5 **Doors:** 4

Rolls Royce Silver Spirit & Spur 1980 - 2000

A Shadow underneath, but on top this is the fabulously well finished wedding car of choice which had a 20 year run. . Spur is 4 inches longer. Updated with 4-speed auto in '91 airbags in '92, plus there was a Silver Dawn a revamped body in '95, then in '97 Spirit dropped; Spur gets light-pressure turbo, 300bhp. Lots of hand-built for zero money. **Engines:** 6.75 **MPG**: 15 **Speed:** 120 **Boot:** 400 **Seats**: 5 **Weight**: 2245 **Doors:** 4

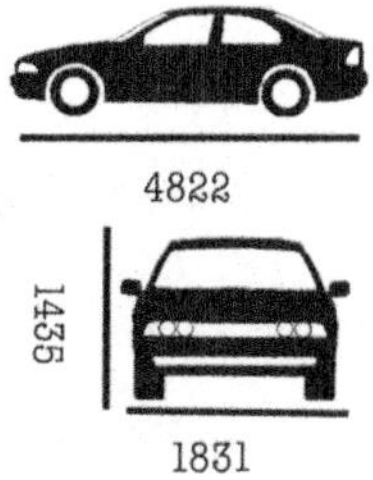

Volvo S80 1998 - 2005

Size matters and they don't come much bigger than the S80. For those who want lots of equipment and a comfortable way to cover long distances this is the perfect vehicle. Underrated and ignored for not being posh enough it is a proper Swedish tank that will rack up massive mileages. Suspension wear, diesel ECUs and general electrics.

Engines: 2.0 • 2.9 • 2.4D **CO2:** 172-232g/km **MPG**: 29.1-43.5
Gearbox: Auto **Boot:** 460 **Seats**: 5 **Doors:** 4

Bangindex by Car Builder

www.freecarmag.com
www.bangernomics.com
www.foresightpublications.com

James Ruppert still regrets ever starting this Bangerpedia and committing to a whole series of these volumes. He has driven 527 of the 528+ models featured here. Always interested in motoring history (grade 1 CSE), he bought all sorts of MOT borderline Bangers that were prehistorically awful. Once sold cars then switched to writing about used ones. Spent many years driving around the country looking at car lots and talking to blokes who sold them. Wrote for newspapers including The Independent, The Sunday Times and Evening Standard. Best known as a writer for Car magazine, Autocar, Performance Car, Supercar Classics and many others. Dreamt up the word Bangernomics in 1990. Founded Free Car Mag the only consumer car magazine launched in the last decade. Believes we are being Demotorized. Banging on about marginal motoring only makes a marginal living. Please consider the following titles for inclusion on your bookshelves to help pay for food, fuel and automotive tat.

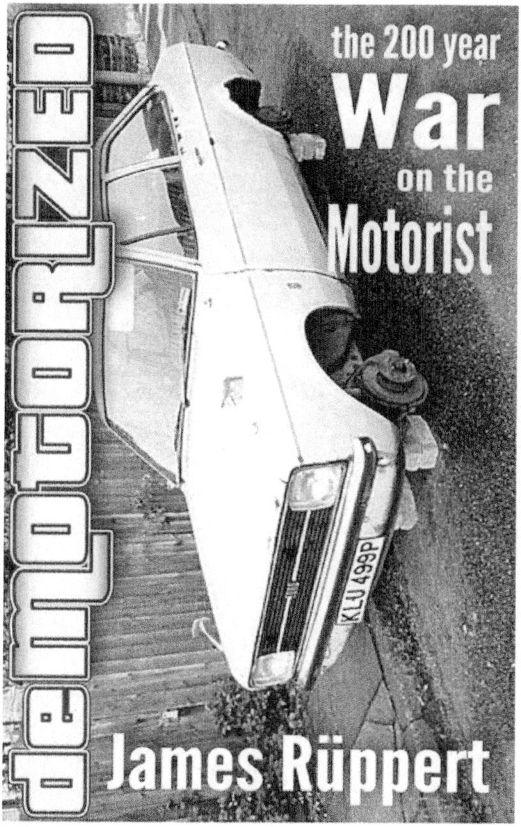

Demotorized. This is the not quite definitive history of how parking became an industry, governments overtaxed motorists, carmakers lied about how safe and environmentally friendly they were and tested their emissions on monkeys and humans. Also why robot cars are so dangerous and electric cars are not the answer. Also, who is behind the ongoing plan to Demotorize the world. Printed in 2020 it predicted the future with terrifying accuracy.
Buy it before it gets banned.

The Award Winning and so good the BBC based a documentary (Das Auto) on it **The German Car Industry, My Part in its Victory.** Winner of the Mercedes-Benz Montagu of Beaulieu Trophy for the best motoring book of year. Contains Yuppies, BMW 3, 5, 6 & 7-Series (yes that's all there was back then) and references to contemporary 1980s music.

The Critically acclaimed and best selling **The British Car Industry Our Part in Its Downfall**, which has been described as "Beautifully written with considerable wit" by the Daily Telegraph and The Daily Express said, "It's informative, well-written and extremely funny" which was really nice of them. The story of the motor industry from 1945 to 2005, explaining just why you can't buy a British built family hatchback from a British owned company. Plus all the cars, James's Dad bought.

Channeling the absurdist spirit of Spike Milligan" said Classic Car, others just called it "Witty", that's **My Mini Cooper Its Part in My Breakdown** which has the full history of the world's favourite small car, plus the history of what came before but not after. Also details the on-off restoration of a Mini Cooper that cost £200 in 1979. The expense of restoring the bodywork and mechanicals ought to put off most sane people. Except that original minis are always worth far more than you could imagine.

Ruppert's Bangerpedia deals with the confusing line up of noughties models. The Bangerpedia does this with a fairly ridiculous rating system involving Slog the Bangernomic Dog. The briefest of guides to each of the 700+ models includes some fairly marginal buying advice, a sarcastic appraisal and an almost definitive indication as to whether it will fit into a garage, or not.

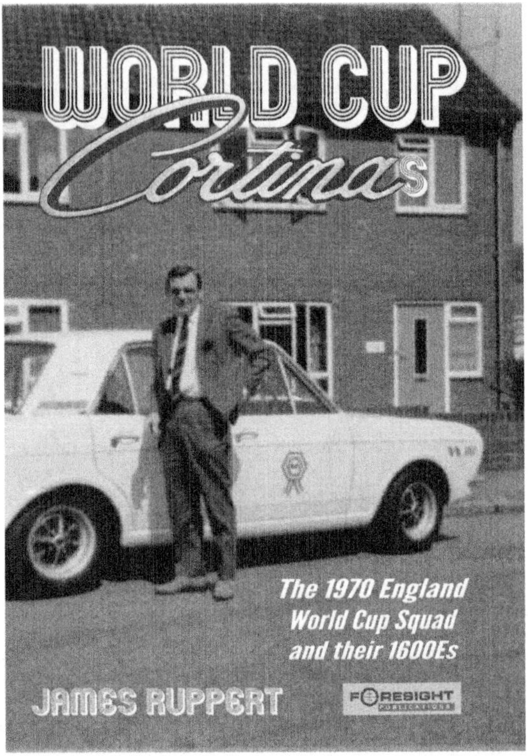

World Cup Cortinas In 1970 something very remarkable happened. Not only had England qualified for the World Cup in Mexico because they were defending champions, but most important of all Ford had loaned every squad member the greatest saloon car in the world. A book about what happened to those Ford Cortinas and how footballers bought better cars when they earned more money and no longer owned the same motors that your mum and dad drove.

Just imagine if BMC had not been so terrible, that Elvis ran a car company, levitating motors were a reality along with steam powered gentleman's carriages and an electricity car designed by and for gentleladies...well, wonder no more. Here is a fully illustrated book which rewrites and redraws automotive history.

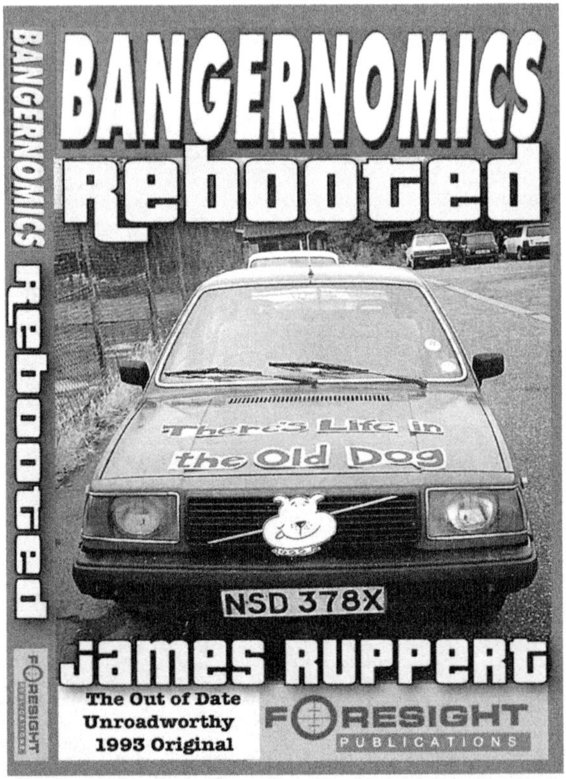

In the beginning, 1993, there was Bangernomics and now there is **Bangernomics Rebooted** which contrasts the absurd expense of buying a new car with the supreme good sense of buying well used. At a stroke depreciation no longer became an issue, running costs were slashed and there were no finance charges to be endured. Here is the original book, republished and effectively rebooted. So it is absolutely useless to any car buyer unless they have access to a time machine.

The world of motoring is in a desperate crisis. Demonised, despised and attacked by every wrong thinking government, local authority, pressure group and even car manufacturers themselves. Bangernomics is here to bring motoring back to the masses, saving time, money and automotive headaches. Squashing together the original Bangernomics, Bangernomics Bible and Bangernomics Diet into one easily digestible guide to buying and running an older cost efficient motor. Only **Bangernomics can save the World.**

There is a new version of the **Bangernomics Bible** coming soon. Not at all sure what the format will actually be. Could be a romantic novel, a pop up might be a lot of bother, maybe a colouring one would be appreciated. Anyway working on it right now...Actually it will be the top 200 Bangernomic tips...

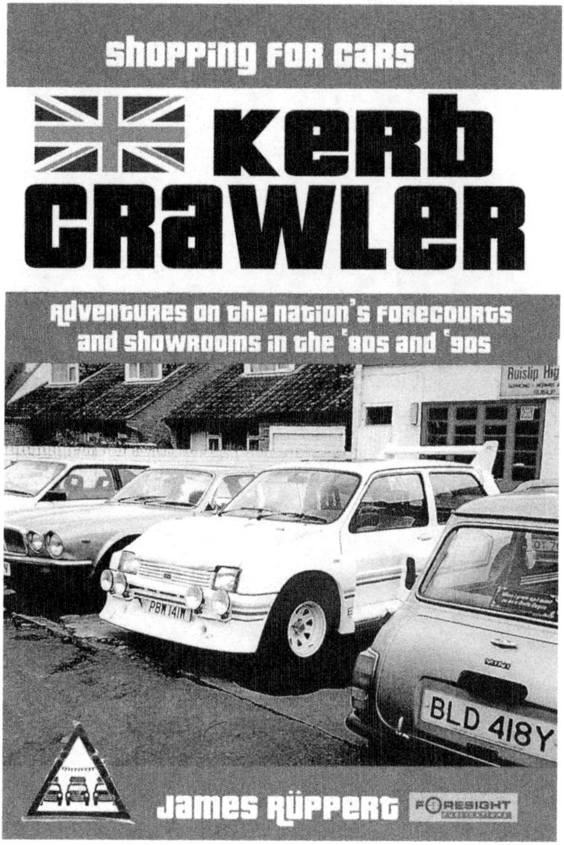

Shopping for cars can be a full time job, well it was for me so here's **Kerbcrawler** In the pre Interweb age it meant actually going to look at cars for sale. It meant talking to the blokes selling the cars face to face. Find out how I got sued by one of the 'stars' of the Italian Job, was told off by the Deputy Prime Minister, find out what the worst car Sir Ian Botham ever owned was, and that time I lent my car to Bjork. A book to dip in and out of...

He was a man running out of time and that time was 1973. **Spencer Haze** is a comic superhero who first hit the 21st century in the year 2000. First on the 4Car website, then later in national newspapers and car magazines. Some say he was even on the television and might well have been real. Here is just about every frame he ever squeezed into, plus a TV script and random pictures that have never been seen before. Here is everything possibly no one ever wanted to know about a fictional bloke called Spencer who loved cars, people and the absolute truth. Here's his **Comicarpocalypse**

Printed in Great Britain
by Amazon

48318601R00178

The only pro-motorist car mag in the Universe*

It costs nothing, but contains everything you need to be a more anarchistic, bolshie and less co-operative car driver.

***www.freecarmag.com**